Editor
Lorin Klistoff, M.A.

Managing Editor
Karen J. Goldfluss, M.S. Ed.

Illustrator
Teacher Created Resources

Cover Artist
Brenda DiAntonis

Creative Director
Karen J. Goldfluss, M.S. Ed.

Art Coordinator
Renée Christine Yates

Art Production Manager
Kevin Barnes

Imaging
Denise Thomas
Nathan Rivera

Publisher
Mary D. Smith, M.S. Ed.

Author

Jo Grinham

(Revised and rewritten by Teacher Created Resources, Inc.)

Teacher Created Resources, Inc.
6421 Industry Way
Westminster, CA 92683
www.teachercreated.com
ISBN-13: 978-1-4206-8994-5

Made in U.S.A.

Teacher Created Resources

Table of Contents

Table of Contents

Introduction

Targeting Math

The series *Targeting Math* is a comprehensive classroom resource. It has been developed so that teachers can find activities and reproducible pages for all areas of the math curriculum.

About This Series

The twelve books in the series cover all aspects of the math curriculum in an easy to access format. Each of the three levels has four books: Numeration and Fractions; Operations and Number Patterns; Geometry, Chance and Data; and Measurement. Each topic in a book is covered by one or more units that are progressive in level. The teacher is able to find resources for all students whatever their ability. This enables the teacher to differentiate for different ability groups. It also provides an easy way to find worksheets at different levels for remediation and extension.

About This Book

Targeting Math: Operations and Number Patterns (Grades 3 and 4) contains the following topics: addition, subtraction, multiplication, division, number facts, and number patterns. Each topic is covered by two complete units of work. (See Table of Contents for specific skills.)

About Each Unit

Each unit is complete in itself. It begins with a list of outcome indicators, resources needed, mathematical language used and a description of each student activity page. This is followed by suggested student activities to reinforce learning. The reproducible pages cover different aspects of the topic in a progressive nature and all answers are included. Every unit includes an assessment page. These assessment pages are important resources in themselves as teachers can use them to find out what their students know about a new topic. They can also be used for assessing specific outcomes when clear feedback is needed.

About the Skills Index

A Skills Index is provided at the end of the book. It lists specific objectives for the student pages of each unit in the book.

ADDITION

These units contain equations in horizontal and vertical algorithms, problem solving, magic squares, number wheels, and money. Skills are also practiced by estimating, calculating totals and working out the perimeter. Students are encouraged to use appropriate mathematical language, differentiate between exact amounts and sensible estimations, and draw on numerous strategies to solve addition tasks. An activity page uses addition in a fun, problem-solving situation. Two assessment pages are included.

ADDITION PRACTICE

Unit 1

Adding to 999
Money
Adding by 10s
Problem solving

Objectives

- check whether answers to problems are correct and sensible
- comprehend and solve written word problems
- carry out operations on money
- recall addition facts to 20
- add numbers with and without regrouping to 999
- write complete equivalent number statements using the = sign
- represent addition facts up to 100 in symbolic form
- use a range of strategies to recall addition facts to 20

Language

ones, tens, hundreds, columns, addition, equals, number sentence, addition square

Materials/Resources

unit cubes (Base 10), blank paper, coins, dice, colored pencils, calculators

Contents of Student Pages

★ Materials needed for each reproducible student page

Page 8 Addition Squares
solving problems, number sentences, addition squares, double-digit addition

Page 9 Wheels and Columns
completing number wheels, single-column addition, and money

★ colored pencils

Page 10 Adding Tens
solving equations and word problems

★ Base 10 materials

Page 11 Regrouping
writing equations, completing an addition chart, and completing number sentences

★ Base 10 materials

Page 12 Assessment

Page 13 Activity—Total 99
adding eight numbers to reach a given total

★ calculators

Remember

Before starting ensure each student:

- ❑ *reads problem solving questions twice.*
- ❑ *checks answers.*

Additional Activities

- *Set up a supermarket with various items priced at different amounts. Play shopping games.*
- *Create giant magic squares.*
- *Have "Speed Races"—teacher calls out various amounts, students keep adding until they cannot keep up. Winner is the student who reaches the final answer.*
- *Create addition problems with dominoes.*
- *Practice addition using dice. Gather three or four players. Take turns to roll four dice. Total the scores. Keep a running total. The first to reach 100 (200, etc.) is the winner.*

Answers

Page 8 Addition Squares

1. a. 16 + 12 + 17 = 45
   b. 15 + 20 + 35 = 70
2.

a.

5	8	13
2	6	8
7	14	21

b.

3	3	6
5	9	14
8	12	20

c.

8	7	15
9	6	15
17	13	30

3. a. Teacher to check.
   b. Teacher to check.
4. a. 105 b. 97
   c. 131 d. 90
   e. 88

Page 9 Wheels and Columns

1.

a.

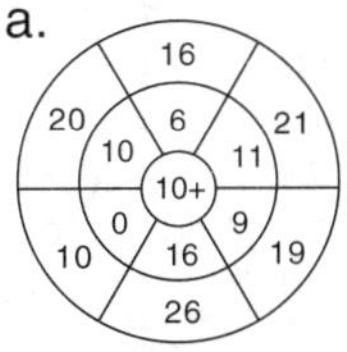

b.

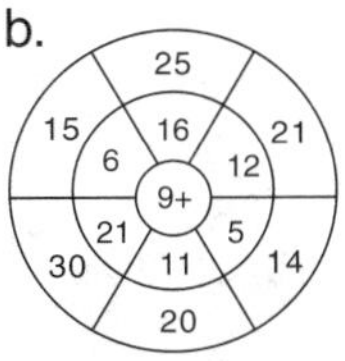

c.

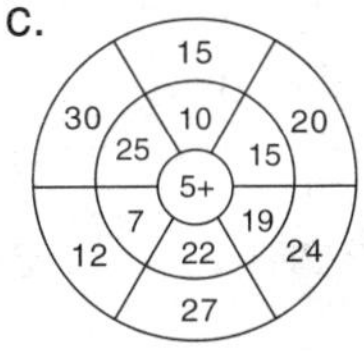

d. Shade blue: (a) 26 (b) 30 (c) 30
e. Shade green: (a) 20, 16, 6, 10, 10, 16, 26 (b) 6, 16, 12, 30, 20,14 (c) 30, 10, 20, 12, 22, 24
f. Circle: (a) 0 (b) 5 (c) 7

2. a. 26 e. 34 h. 23
   b. 18 f. 24 i. 29
   c. 18 g. 32 j. 38
   d. 24
3. a. $5.00 d. $7.85 g. $5.95
   b. $2.50 e. $2.20 h. $10.60
   c. $4.95 f. $4.55

Page 10 Adding Tens

1. a. 30 d. 80 g. 130
   b. 40 e. 110 h. 100
   c. 60 f. 110
2. a. 120 d. 90 g. 140
   b. 180 e. 180 h. 110
   c. 60 f. 130
3. Shade a and e
4. a. 110 cards
   b. 220 stamps

Page 11 Regrouping

1. a. 80 e. 82
   b. 116 f. 92
   c. 82 g. 93
   d. 82 h. 66
2. 16 21 15 28 37 53 52
   27 32 26 39 48 64 63
   29 34 28 41 50 66 65
   34 39 33 46 55 71 70
3. a. 473 e. 939 i. 494
   b. 861 f. 269 j. 393
   c. 546 g. 775 k. 297
   d. 364 h. 663 l. 800

Page 12 Assessment

1. a. 14 d. 21 g. 43
   b. 29 e. 33 h. 26
   c. 18 f. 60 i. 42
2. a. 103 e. 105 h. $7.01
   b. 116 f. $3.65 i. $8.34
   c. 125 g. $4.17 j. $11.00
   d. 151
3.

a.

30	5	35
6	4	10
36	9	45

b.

38	6	44
16	22	38
54	28	82

c. Answers will vary.

4. a. 140 d. 160 g. 200
   b. 80 e. 250 h. 240
   c. 70 f. 90 k. 150

Page 13 Activity—Total 99

1. 3, 19, 7, 8, 15, 40, 5, 2
2. 7, 15, 15, 16, 9, 30, 3, 4
3. 16, 12, 8, 11, 1, 32, 9, 10
4. Answers will vary.

Name **Date**

1. Solve the addition problems by writing them out as number sentences.

a. During a sale, Tom sold 16 cars on Tuesday, 12 cars on Wednesday, and 17 cars on Thursday. How many cars did he sell altogether?

_____ + _____ + _____ = _____

b. Ben went to the fruit shop and bought an apple for 15¢, a banana for 20¢, and a peach for 35¢. How much did he spend?

_____ + _____ + _____ = _____

a.

	Tens	Ones
+		

b.

	Tens	Ones
+		

2. Complete the addition squares.

a.

	8	13
2	6	
7		21

b.

	3	6
5		14
	12	20

c.

8		15
	6	15
17		

3. Make your own addition squares and give them to a friend to check.

a.

b.

Remember

The first 2 numbers add to the third number in each row and column.

4. Make sure you show all your work.

a. $\begin{array}{r} 67 \\ +\ 38 \\ \hline \end{array}$ **b.** $\begin{array}{r} 58 \\ +\ 39 \\ \hline \end{array}$ **c.** $\begin{array}{r} 44 \\ +\ 87 \\ \hline \end{array}$ **d.** $\begin{array}{r} 83 \\ +\ 7 \\ \hline \end{array}$ **e.** $\begin{array}{r} 56 \\ +\ 32 \\ \hline \end{array}$

Name **Date**

1. Complete each number wheel.

a.

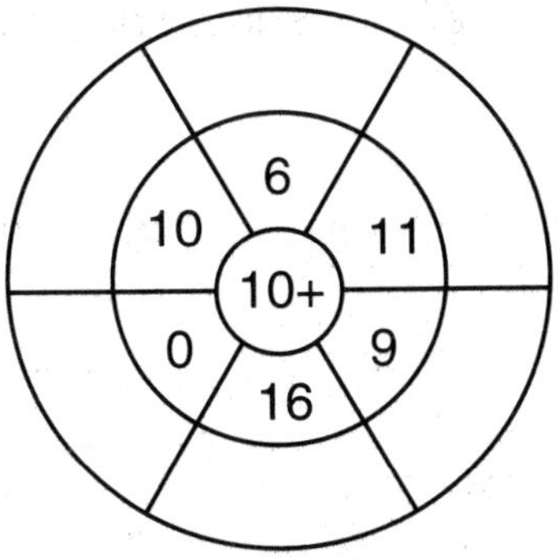

b.

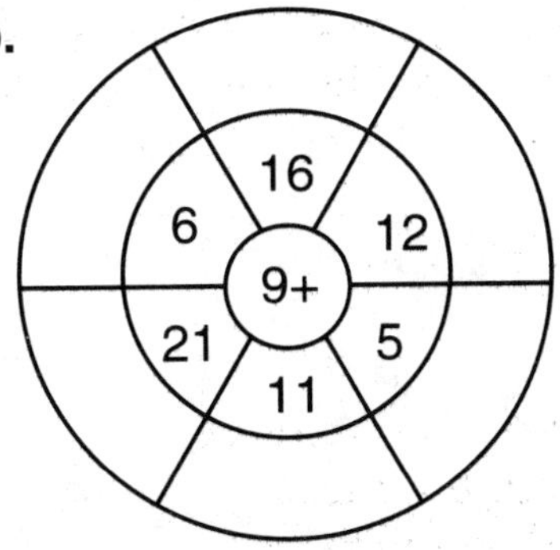

c.

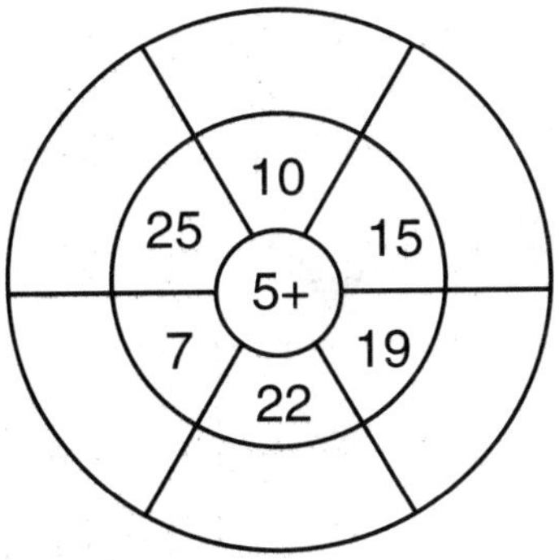

d. Shade the largest number on each wheel blue.

e. Shade the even numbers green on each wheel.

f. Circle the smallest number on each wheel.

2. Remember the ones and tens columns.

a.	**b.**	**c.**	**d.**	**e.**
9	8	6	7	9
8	7	5	8	8
6	2	4	6	9
+ 3	+ 1	+ 3	+ 3	+ 8

f.	**g.**	**h.**	**i.**	**j.**
3	10	13	15	17
4	7	7	5	7
8	9	2	6	7
+ 9	+ 6	+ 1	+ 3	+ 7

3.

a.	**b.**	**c.**	**d.**
\$3.00	\$1.50	\$3.55	\$5.60
+ \$2.00	+ \$1.00	+ \$1.40	+ \$2.25

e.	**f.**	**g.**	**h.**
\$1.10	\$2.95	\$3.75	\$7.70
+ \$1.10	+ \$1.60	+ \$2.20	+ \$2.90

Name **Date**

1. Add these tens.

   a. 20 + 10 = _____ b. 20 + 20 = _____ c. 30 + 30 = _____ d. 40 + 40 = _____

   e. 30 + 80 = _____ f. 90 + 20 = _____ g. 70 + 60 = _____ h. 50 + 50 = _____

2. Add these numbers using the columns.

a.	b.	c.	d.
70	80	20	30
30	90	20	40
+ 20	+ 10	+ 20	+ 20

e.	f.	g.	h.
90	10	80	20
80	30	50	10
+ 10	+ 90	+ 10	+ 80

3. Shade the balloons that add to a total of 100.

   a. b. c. d. e.

4. Add these groups of tens to solve the problems. Use Base 10 blocks to help.

   a. Patty had 50 baseball cards, 30 football cards, 10 basketball cards, and 20 golf cards. How many cards did she have altogether?

   ______ + ______ + ______ + ______ = ______

   b. Bob collected 70 stamps in 1996, 80 stamps in 1997, 30 stamps in 1998, and 40 stamps in 1999. How many stamps did he have altogether?

   ______ + ______ + ______ + ______ = ______

Name **Date**

1. Write the problem in the boxes, then complete the number sentence. Use Base 10 materials to model each operation to help you.

a. 48 + 32 = _____ **b.** 72 + 44 = _____ **c.** 63 + 19 = _____ **d.** 33 + 49 = _____

	4	8
+	3	2

e. 15 + 67 = _____ **f.** 69 + 23 = _____ **g.** 54 + 39 = _____ **h.** 37 + 29 = _____

2. Complete the addition chart.

	+	10	15	9	22	31	47	46
a.	6							
b.	17							
c.	19							
d.	24							

3. Write the number.

a. 400 + 70 + 3 = _________ **b.** 850 + 10 + 1 = _________ **c.** 520 + 20 + 6 = _________

d. 300 + 60 + 4 = _________ **e.** 900 + 30 + 9 = _________ **f.** 240 + 26 + 3 = _________

g. 700 + 73 + 2 = _________ **h.** 630 + 30 + 3 = _________ **i.** 430 + 58 + 6 = _________

j. 330 + 54 + 9 = _________ **k.** 270 + 20 + 7 = _________ **l.** 770 + 25 + 5 = _________

Name **Date**

1. Complete.

a. 6 + 8 = ______ b. 17 + 12 = ______ c. 9 + 9 = ______

d. 10 + 11 = ______ e. 20 + 13 = ______ f. 20 + 20 + 20 = ______

g. 30 + 7 + 6 = ______ h. 4 + 12 + 10 = ______ i. 7 + 30 + 5 = ______

2. Show all work.

a.	b.	c.	d.	e.
64 + 39	54 + 62	79 + 46	84 + 67	56 + 49

f.	g.	h.	i.	j.
\$2.05 + \$1.60	\$2.32 + \$1.85	\$4.84 + \$2.17	\$5.61 + \$2.73	\$5.00 + \$6.00

3. Complete these addition squares.

a.

30	5	
	4	10
36		45

b.

	6	44
16		38
54	28	

c.

45		
	15	

4. a. 60 + 70 + 10 = ______ b. 50 + 20 + 10 = ______ c. 20 + 20 + 30 = ______

d. 30 + 50 + 80 = ______ e. 90 + 90 + 70 = ______ f. 60 + 10 + 20 = ______

g. 50 + 90 + 60 = ______ h. 70 + 70 + 100 = ______ i. 100 + 20 + 30 = ______

Name **Date**

Look at the groups of numbers. Circle the 8 numbers that total 99. You can use a calculator.

1.

3 + 9 + 19 + 7 + 6 + 8 + 15
+ 40 + 20 + 5 + 13 + 2 = 99

2.

7 + 15 + 15 + 16 + 21 + 9 +
4 + 30 + 3 + 4 + 18 + 8 +
11 + 5 = 99

3.

16 + 12 + 8 + 3 + 2 + 11 +
26 + 50 + 1 + 32 + 5 + 9 +
10 = 99

4. Design a similar problem of your own. Ask a friend to solve it.

MORE ADDITION PRACTICE

Unit 2

Adding to 999
Money
Number wheels
Perimeter
Problem solving

Objectives

- identify ones, tens, and hundreds
- add numbers with and without regrouping to 999
- carry out operations on money
- check whether answers to problems are correct and sensible
- measure and record perimeter
- interpret common spatial language

Language

ones, tens, hundreds, working out, number sentence, estimate, exact, less than, greater than, perimeter, total

Materials/Resources

unit cubes (Base 10), blank paper, paper money (optional), colored pencils, rulers

Contents of Student Pages

** Materials needed for each reproducible student page*

Remember

Before starting ensure each student:

- ❑ *knows the importance of showing all work.*
- ❑ *understands the importance of working neatly to eliminate careless mistakes.*
- ❑ *realizes the importance of checking all answers.*

Additional Activities

- ❑ *Students explore school playground and perimeters of boundaries. Draw areas to scale.*
- ❑ *Students find distances to various destinations and calculate round trips from the school gates.*
- ❑ *Students bring in advertising brochures. Have students go on a make-believe shopping spree keeping a total of the money they spend.*
- ❑ *Make an addition crossword.*

Answers

Page 16 Number Calculating

1. a. 851 b. 995 c. 1,210 d. 872 e. 685 f. 985 g. 681 h. 1,278
2. a. < b. < c. < d. > e. > f. <
3. a.

32	17	20
11	23	35
26	29	14

b.

31	21	44
45	32	19
20	43	33

c.

20	30	28
34	26	18
24	22	32

4. Answers will vary.

Page 17 Money

1. a. \$440 + \$630 = \$1,070
   b. \$400 + \$630 = \$1,030
   c. \$800 + \$90 = \$890
   d. \$550 + \$700 = \$1,250
2. a.

\$		4	3	5
+		6	3	4
\$	1,	0	6	9

b.

\$		4	0	0
+		6	3	1
\$	1,	0	3	1

c.

\$		7	9	9
+			8	9
\$		8	8	8

d.

\$		5	4	5
+		6	9	9
\$	1,	2	4	4

3. a. \$64.95 + \$73.00 + \$45.00 + \$7.85 = \$190.80
   b. \$16.95 + \$16.95 + \$16.95 = \$50.85

Page 18 Totals

1. a. 64 b. 48 c. 49 d. 43 e. 65 f. 49 g. Ashley h. Lucy
2. a.

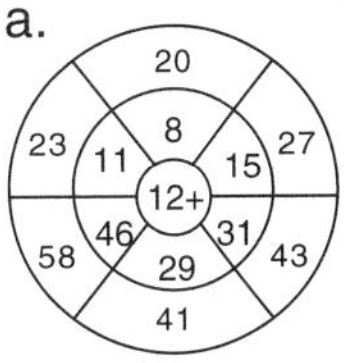

b.

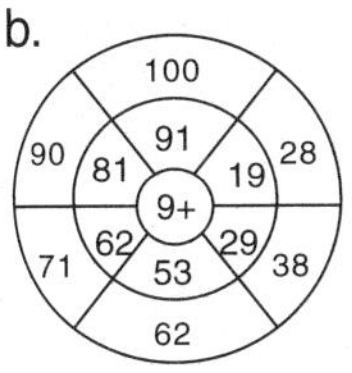

c.

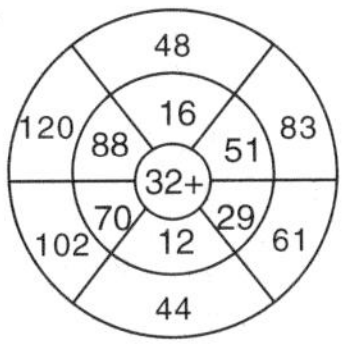

   d. 120 located in letter "c" wheel should be colored green.
   e. 20 located in letter "a" wheel should have a cross over it.
   f. The following numbers should be colored yellow in the number wheels: 20, 58, 90, 100, 28, 38, 62, 120, 102, 44, and 48.
3. a. 117 b. 63 c. 116 d. 137 e. 130 f. 186 g. 130 h. 116 i. 141 j. 112

Page 19 Perimeter

1. a. 142 m
   b. 140 m
   c. 131 cm
   d. 215 cm
   e. 160 m
   f. Shade in the shape in letter "d" blue.
   g. Shade in the shape in letter "c" red.
2. a. \$11.00
   b. \$8.62
   c. \$8.75
   d. \$7.10
   e. \$4.20
   f. \$11.55
   g. \$14.00
   h. \$5.65

Page 20 Assessment

1. a. 122
   b. 419
   c. 1,131
   d. 995
   e. 810
2. \$95.00 + \$17.50 + \$250.00 + \$150.65 = \$513.15
3. 64
4. a.

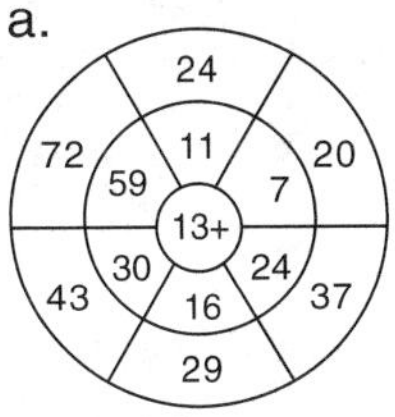

b.

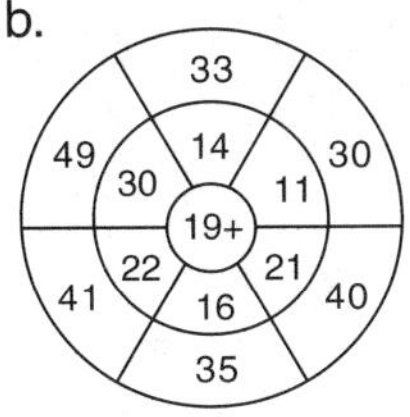

5. a. 148 in.
   b. 60 cm
   c. 98 in.

Name **Date**

1. Find the answers to these number problems.

a. 637 + 214	**b.** 432 + 563	**c.** 536 + 674	**d.** 545 + 327
e. 399 + 286	**f.** 789 + 196	**g.** 222 + 459	**h.** 639 + 639

2. Write < or > to make the statement true.

a. 23 + 24 ☐ 16 + 32
b. 15 + 18 ☐ 19 + 22
c. 23 + 27 ☐ 25 + 26
d. 32 + 56 ☐ 54 + 31
e. 54 + 26 ☐ 40 + 39
f. 62 + 25 ☐ 32 + 60

< is less than
> is greater than

3. Complete these magic squares.

a.

32		
26	29	

69

b.

31		44
	32	
20		

96

c.

		28
24		32

78

4. Make your own magic squares with numbers greater than 20.

32		
26	29	

☐

31		44
	32	
20		

☐

☐

Name **Date**

1. Estimate the total cost of the two items in each question. Round each price to the nearest ten before adding.

   a. Microwave $435 and Stove $634

   Estimation = ________________

   b. TV $400 and Fridge $631

   Estimation = ________________

   c. Video $799 and Film $89

   Estimation = ________________

   d. Washing machine $545, Dryer $699

   Estimation = ________________

2. Now work out the exact answer for each question. Compare this total with your estimate.

   a. +

   b. +

   c. +

   d. +

3. **a.** James went shopping and bought a coat for $64.95, a pair of trousers for $73.00, a pair of shoes for $45.00, and a pair of socks for $7.85. How much did he spend?

   b. Deanne bought her daughter, Brittany, three new pairs of pants at $16.95 each. How much did she spend?

Name **Date**

1. Find the total scores for each child.

	Child	Scores	Total
a.	Isabella	12, 17, 35	
b.	Harry	16, 16, 16	
c.	James	11, 18, 20	
d.	Lucy	20, 4, 19	
e.	Ashley	21, 15, 29	
f.	Courtney	28, 15, 6	

g. Whose total is the highest? ____________________

h. Whose total is the lowest? ____________________

2. Complete each number wheel.

a.

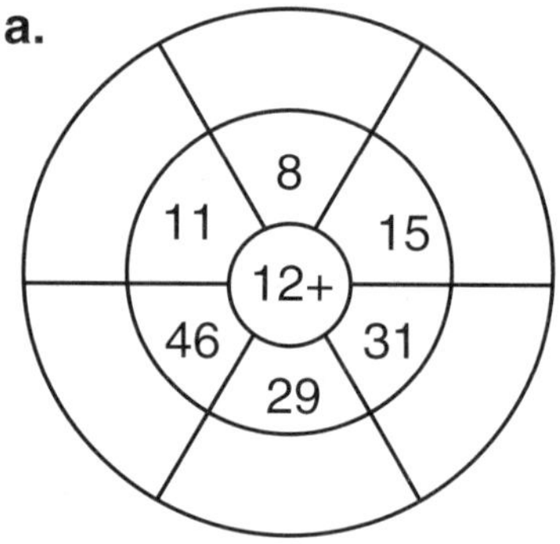

b.

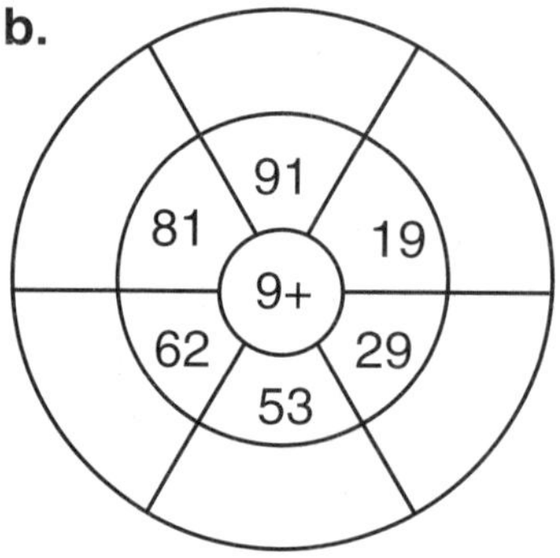

c.

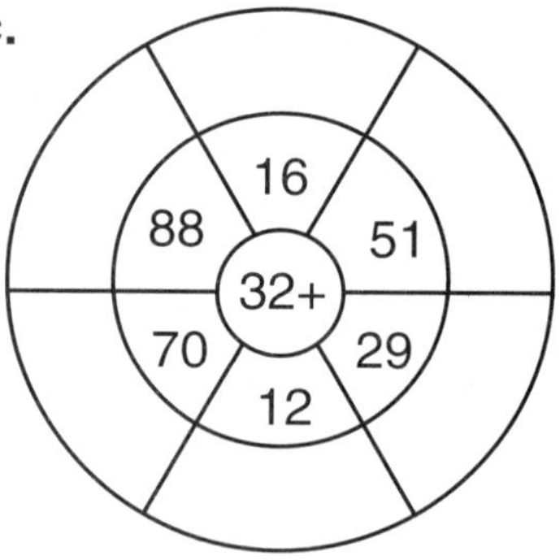

d. Shade the largest number green.

e. Put a cross over the smallest sum.

f. Color all the even answers yellow.

3. Find the answers. Watch the columns when you work out each problem below.

a. $63 + 54$

b. $29 + 34$

c. $55 + 61$

d. $79 + 58$

e. $83 + 47$

f. $99 + 87$

g. $60 + 70$

h. $33 + 83$

i. $72 + 69$

j. $83 + 29$

Name **Date**

1. Find the perimeter of each shape. Remember that the perimeter of a shape is the distance around the outside of the shape.

a.

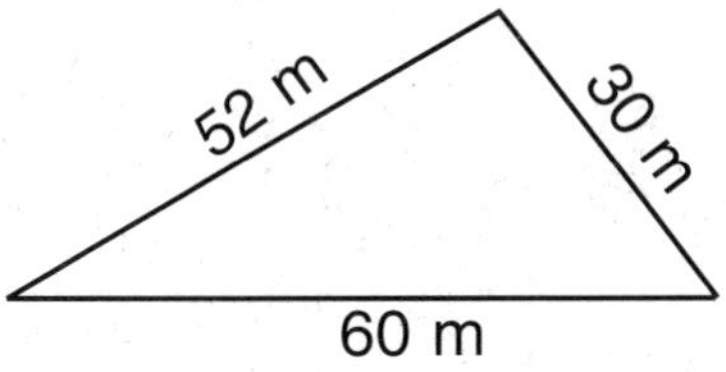

b.

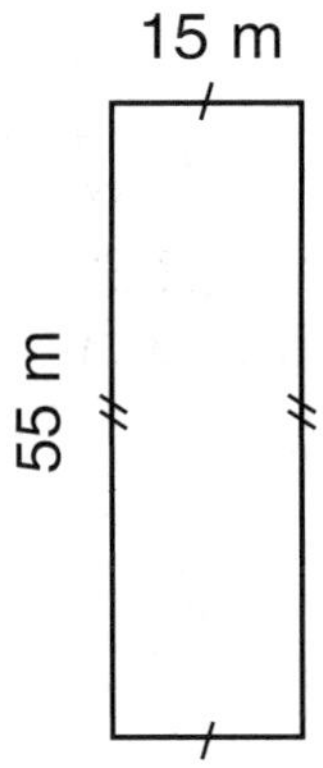

c.

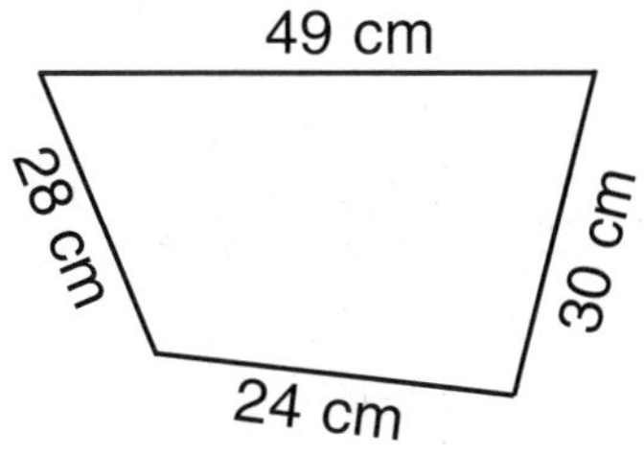

d.

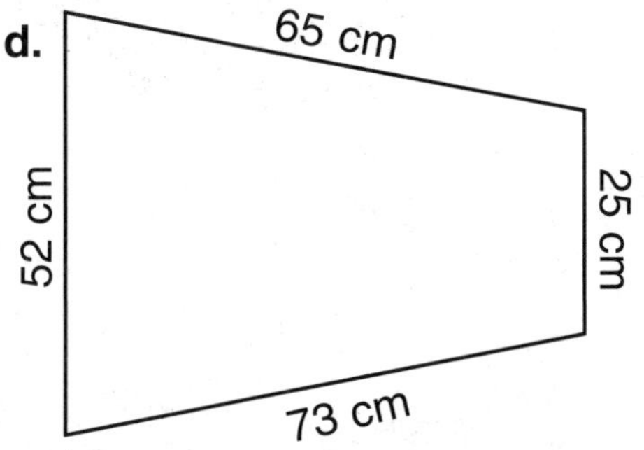

e.

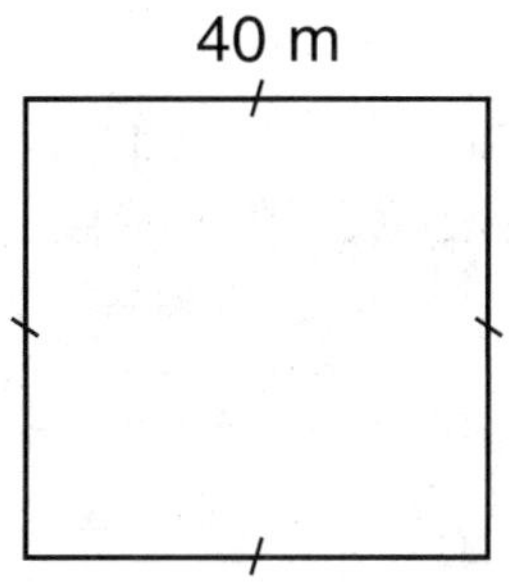

Perimeters:

a. ______________________ **b.** ______________________

c. ______________________ **d.** ______________________

e. ______________________

f. Shade the shape with the greatest perimeter blue.

g. Shade the shape with the smallest perimeter red.

2.

a. $6.05 + $4.95

b. $5.73 + $2.89

c. $6.75 + $2.00

d. $4.35 + $2.75

e. $0.75 + $3.45

f. $8.90 + $2.65

g. $7.37 + $6.63

h. $4.00 + $1.65

Name **Date**

1.

a. 55	**b.** 352	**c.** 631	**d.** 785	**e.** 333
+ 67	+ 67	+ 500	+ 210	+ 477

2. Bob went to pick his Porsche up from the mechanic. A new muffler cost $95.00, the wiper blades cost $17.50, the new tires cost $250.00, and the new leather steering wheel cost $150.65. Show your work below to find out how much it all cost.

3. Sam scored 15 in the math test, Campbell scored 20, Brittany scored 18, and Jarrod scored 11. What was the total of the scores? ____________________

4. Complete these addition wheels.

a.

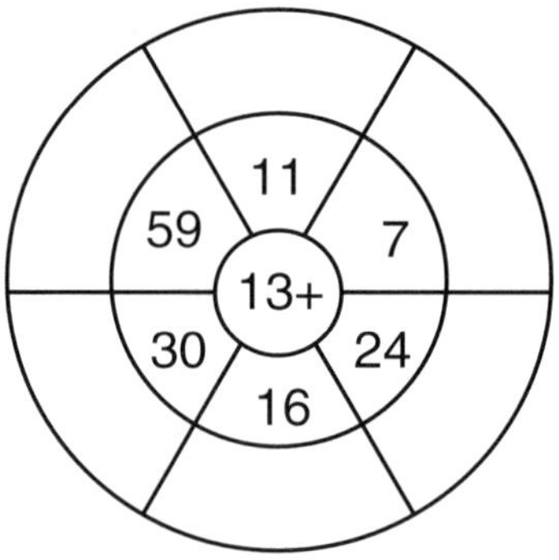

b.

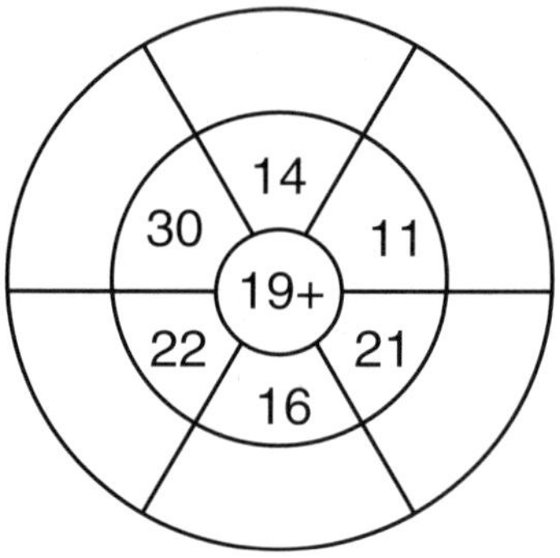

5. Find the total perimeter of these three shapes.

a.

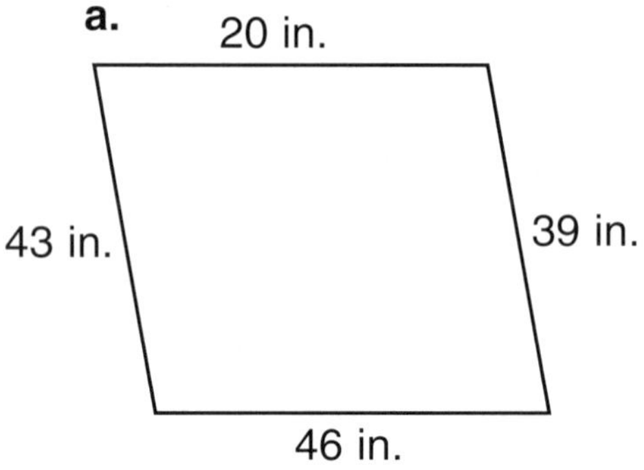

b.

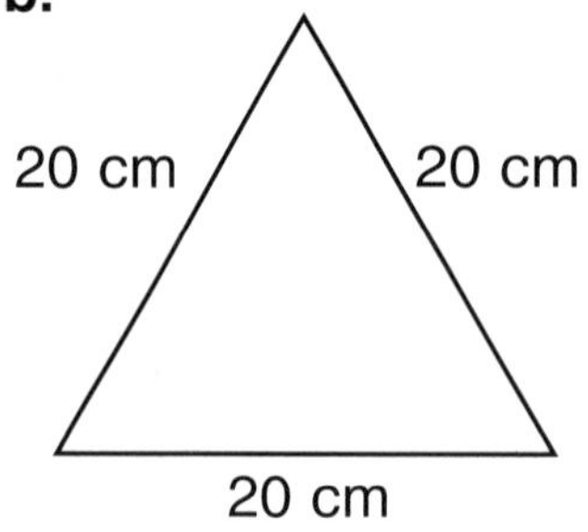

c.

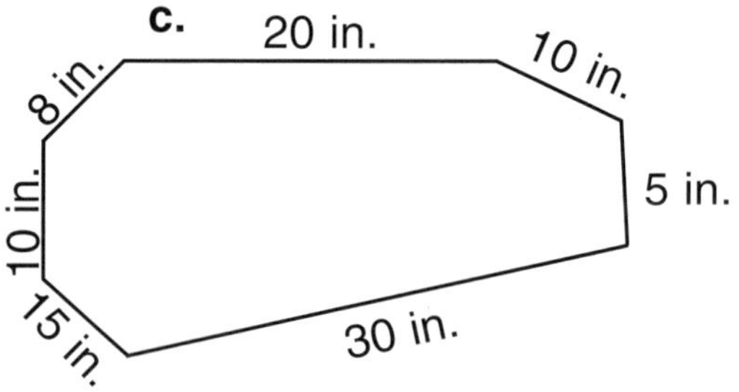

Perimeter = __________ Perimeter = __________ Perimeter = __________

SUBTRACTION

These units contain exercises in subtraction using Base 10 materials, counters, picture number sentences, vertical and horizontal algorithms, and problem solving. Activities—including magic squares, number wheels, subtraction tables, money algorithms, and number patterns—all use subtraction skills. Estimation is encouraged as is the appropriate use of mathematical language. There is some calculator work. The activity page contains a subtraction game and a fun word puzzle to solve. Two assessment pages are included.

SUBTRACTION PRACTICE

Unit 1

Using concrete materials
Problem solving
Regrouping

Objectives

- *represent subtraction as taking an object or objects from a group of objects*
- *write equivalent number sentences using the = sign*
- *comprehend written word problems and transfer them to number problems*
- *identify units and tens*
- *use one or more strategies to solve mathematical problems*
- *subtract numbers using the regrouping method*
- *comprehend and solve written word problems*

Language

ones, tens, hundreds, columns, subtraction, taking away, regrouping, equals, number sentence, odd, even

Materials/Resources

unit cubes (Base 10), blank paper, dice, colored pencils, counters, craft sticks

Contents of Student Pages

* *Materials needed for each reproducible student page*

Page 24 Solve It!
using concrete materials for problem solving, number sentences, and recording
* *unit cubes, craft sticks, counters, pencils*

Page 25 What's Missing?
solving number wheels, missing numbers, and word problems
* *colored pencils*

Page 26 Complete It!
solving equations, number squares, and subtracting tens
* *Base 10 blocks*

Page 27 Regrouping
using base 10 blocks to solve problems, solving equations and word problems
* *Base 10 blocks*

Page 28 Assessment
* *Base 10 blocks*

Page 29 Activity—Puzzle and Game
solving a puzzle using subtraction words and playing a subtraction game for 2–4 players
* *dice, scratch paper*

Remember

Before starting a student:

- ❑ *understands that spacing between each question is important.*
- ❑ *understands that neatness and setting out can affect the final answer.*
- ❑ *reads problem solving questions twice.*
- ❑ *always reads subtraction algorithms from the top.*

Additional Activities

- ❑ *Play this game for three or four players. Each player starts with a score of 100 (200, etc.). Take turns rolling dice and subtracting the number rolled from 100. The first to reach 0 is the winner.*
- ❑ *Play shopping games concentrating particularly on the "change" given to the customer.*
- ❑ *Create giant magic subtraction squares.*
- ❑ *Play a game of subtraction dominoes.*
- ❑ *Find a map with distances between places marked. Work out distances from A to B using the subtraction method.*

Answers

Page 24 Solve It!

1. a. $\begin{array}{r}23\\-22\\\hline 1\end{array}$ d. $\begin{array}{r}20\\-14\\\hline 6\end{array}$

   b. $\begin{array}{r}25\\-3\\\hline 22\end{array}$ e. $\begin{array}{r}40\\-6\\\hline 34\end{array}$

   c. $\begin{array}{r}40\\-10\\\hline 30\end{array}$ f. $\begin{array}{r}18\\-13\\\hline 5\end{array}$

2. a. 53 – 30 = 23
   b. 60 – 43 = 17
   c. 45 – 27 = 18
3. a. 5 – 2 – 2 = 1
   b. 9 – 3 – 4 = 2
   c. 8 – 3 – 4 = 1
   d. 7 – 5 – 2 = 0

Page 25 What's Missing?

1. a.

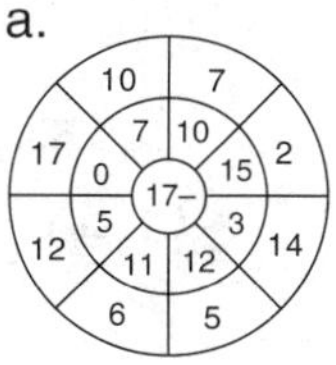

b.

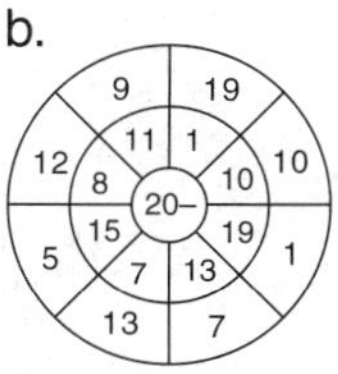

c.

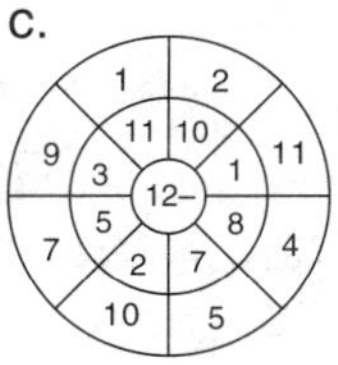

   d. The following numbers on the number wheels should be shaded red: letter "a" wheel—10, 2, 14, 6, 12; letter "b" wheel—12, 10; letter "c" wheel—2, 4, 10
   e. The following numbers on the number wheels should be shaded yellow: letter "a" wheel—17, 7, 5; letter "b" wheel—7, 19, 1, 7, 13, 5; letter "c" wheel—9, 1, 11, 5, 7

2. a. 9 e. 11 i. 19
   b. 11 f. 14 j. 12
   c. 3 g. 19 k. 12
   d. 4 h. 3 l. 24
3. a. 16 – 9 = 7
   b. 20 – 16 = 4
   c. 30 – 6 = 24

Page 26 Complete It!

1. a. $\begin{array}{r}50\\-20\\\hline 30\end{array}$ b. $\begin{array}{r}40\\-20\\\hline 20\end{array}$ c. $\begin{array}{r}30\\-10\\\hline 20\end{array}$
2. a. 20 d. 20
   b. 20 e. 50
   c. 20
3. a. (–) ↓

12	5	7
8	4	4
4	1	3

b. (–)

20	13	7
6	4	2
14	9	5

c. (–)

14	8	6
5	1	4
9	7	2

4. Answers will vary. Check for accuracy.
5. a. 32 d. 13 g. 52
   b. 27 e. 22 h. 43
   c. 42 f. 23

Page 27 Regrouping

1. a. $\begin{array}{r}33\\-26\\\hline 7\end{array}$ d. $\begin{array}{r}14\\-8\\\hline 6\end{array}$

   b. $\begin{array}{r}42\\-26\\\hline 16\end{array}$ e. $\begin{array}{r}36\\-18\\\hline 18\end{array}$

   c. $\begin{array}{r}23\\-15\\\hline 8\end{array}$ f. $\begin{array}{r}50\\-23\\\hline 27\end{array}$

2. a. 17 f. 16 k. 34
   b. 13 g. 25 l. 23
   c. 7 h. 44 m. 13
   d. 27 i. 28 n. 24
   e. 33 j. 7 o. 36
3. a. 23 – 16 = 7 $\begin{array}{r}23\\-16\\\hline 7\end{array}$ b. 65¢ – 38¢ = 27¢ $\begin{array}{r}65¢\\-38¢\\\hline 27¢\end{array}$

Page 28 Assessment

1. a. $\begin{array}{r}35\\-13\\\hline 22\end{array}$ b. $\begin{array}{r}43\\-11\\\hline 32\end{array}$
2. a.

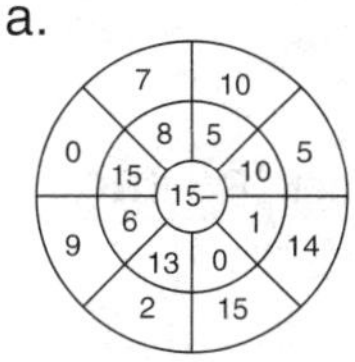

b.

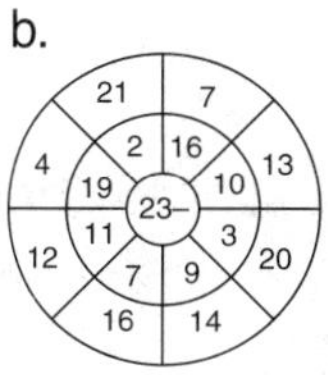

c.

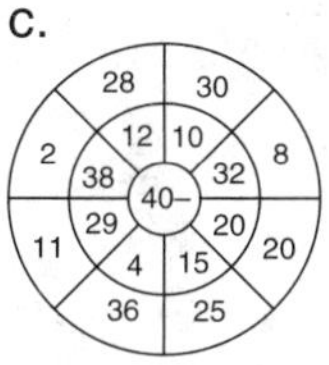

3. a. 6 d. 13 g. 28
   b. 12 e. 18 h. 59
   c. 23 f. 39
4. a. 21 cars b. $13.00 c. 38 books

Page 29 Activity—Puzzle and Game

1. FIFTY minus THREE equals FORTY-SEVEN
2. THIRTY-NINE minus NINE equals THIRTY
3. TWENTY minus SIX equals FOURTEEN
4. SIXTY-ONE take away EIGHT equals FIFTY-THREE
5. Answers will vary. Check for accuracy.

Name ____________________ **Date** ____________

1. Count, then write and solve the problem.

a. 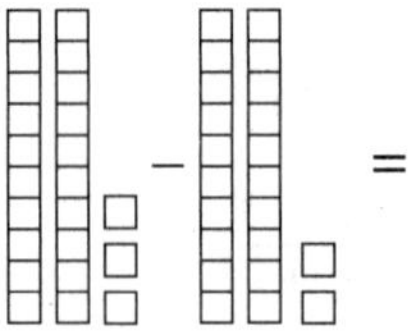

Tens	Ones
–	

b. 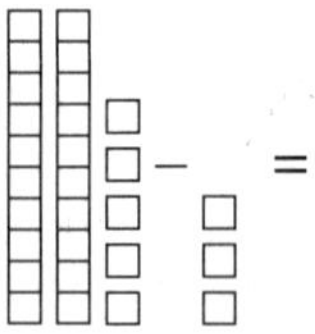

Tens	Ones
–	

c. 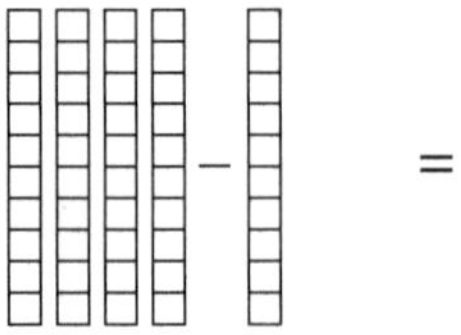

Tens	Ones
–	

d. 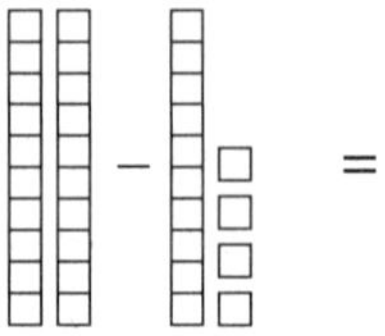

Tens	Ones
–	

e. 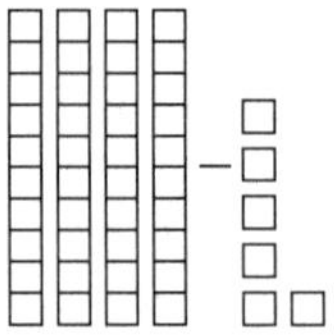

Tens	Ones
–	

f. 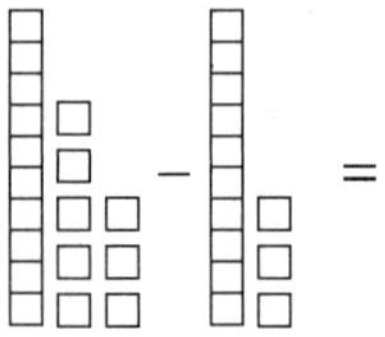

Tens	Ones
–	

2. Use craft sticks, pencils, or counters to model and help solve the problems below.

a. Mr. Blue had 53 plants to sell. He sold 30 during the week. How many did he have left over?

_____ – _____ = _____

b. Sixty apples were delivered to the fruit shop on Monday. By Thursday, 43 had been sold. How many were left in the shop?

_____ – _____ = _____

c. There were 45 cars in the race. Twenty-seven finished in record time. How many cars were still to go over the finish line?

_____ – _____ = _____

3. Write a number sentence to match the pictures.

a. 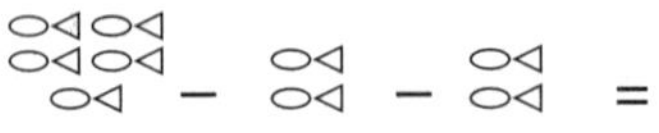

_____ – _____ – _____ = _____

b. 

_____ – _____ – _____ = _____

c. 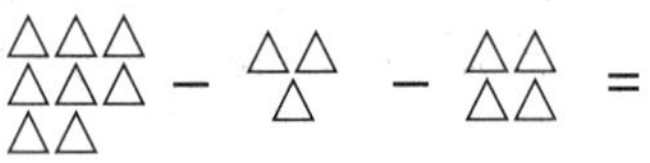

_____ – _____ – _____ = _____

d. 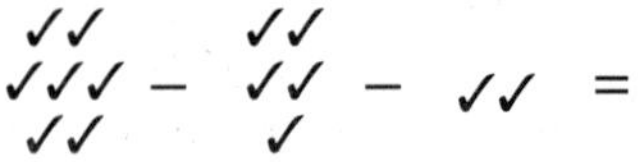

_____ – _____ – _____ = _____

Name **Date**

1. Complete each number wheel.

a.

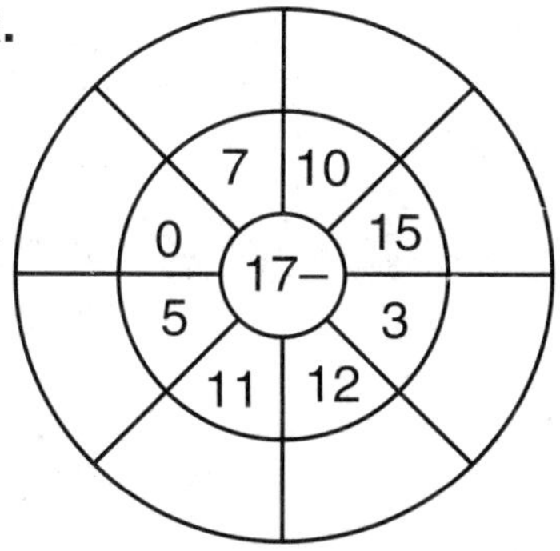

b.

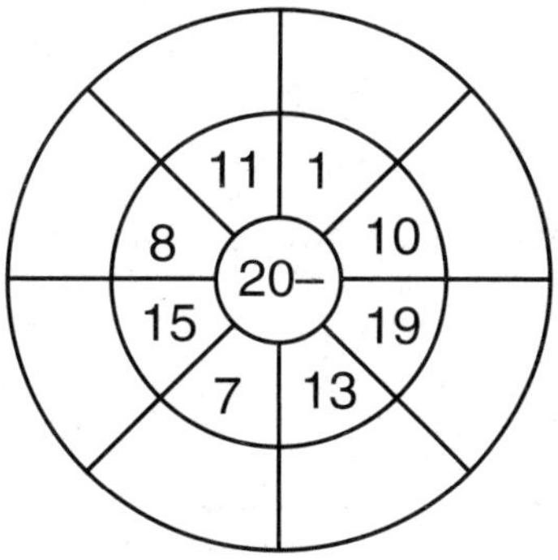

c.

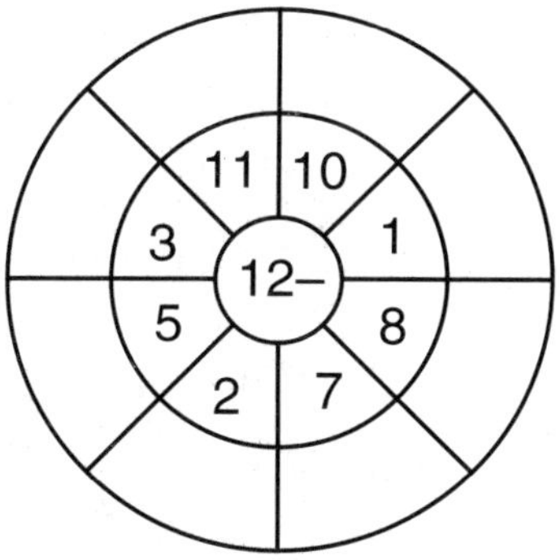

d. Shade all the even answers red.

e. Shade all the odd answers yellow.

2. Find the missing number in each question.

a. 18 − 9 = ___

b. 22 − 11 = ___

c. 13 − ___ = 10

d. 19 − 15 = ___

e. ___ − 4 = 7

f. 19 − ___ = 5

g. ___ − 11 = 8

h. 19 − ___ = 16

i. 22 − 3 = ___

j. 21 − ___ = 9

k. 30 − ___ = 18

l. ___ − 9 = 15

3. Solve the subtraction problems by writing them out as number sentences.

a. At his birthday party Tom received 16 gifts. He had opened 9 of the gifts. How many did he have left?

_____ – _____ = _____

b. Tom had invited 20 friends, but only 16 came. How many children were unable to come to the party?

_____ – _____ = _____

c. Tom's favorite party food was chocolate cookies. His mom made 30 chocolate cookies. When everyone had gone home, Tom found 6 chocolate cookies. How many were eaten at the party?

_____ – _____ = _____

Name **Date**

1. Subtract groups of 10.

a. 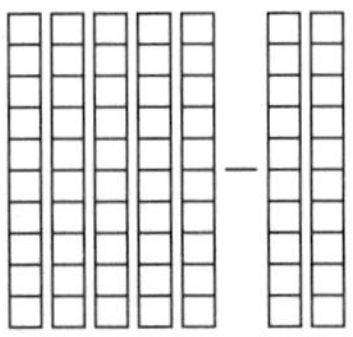

Tens	Ones

b. 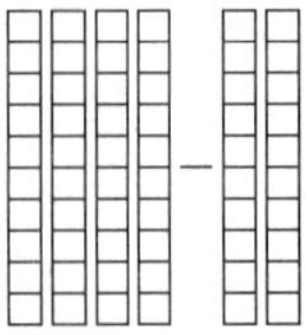

Tens	Ones

c. 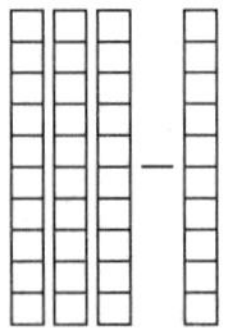

Tens	Ones

2. Complete these problems. Watch the columns!

a. 80 − 60

b. 50 − 30

c. 90 − 70

d. 40 − 20

e. 60 − 10

3. Complete the subtraction squares. Subtract rows and columns (not diagonals).

a. (−)

12	5	7
8	4	
		3

b. (−)

	13	7
6	4	
14		5

c. (−)

14	8	
	1	4
9		2

4. Make your own subtraction squares. Give them to a friend to check.

a. (−)

b. (−)

c. (−)

5. Complete these problems. Watch the columns!

a. 63 − 31

b. 59 − 32

c. 64 − 22

d. 29 − 16

e. 44 − 22

f. 57 − 34

g. 69 − 17

h. 75 − 32

Name **Date**

1. Complete the number sentences. Model each problem where needed.

a. 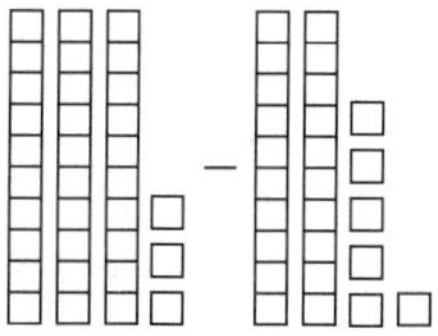 – =

Tens	Ones
–	

b. 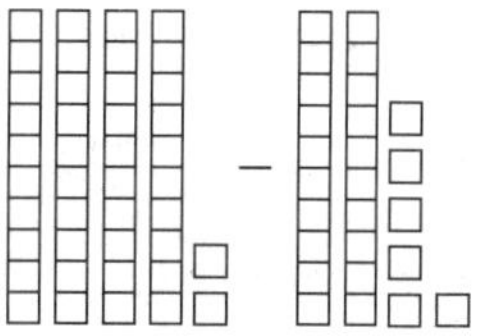 – =

Tens	Ones
–	

c. 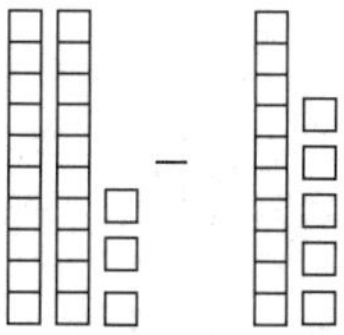 – =

Tens	Ones
–	

d. 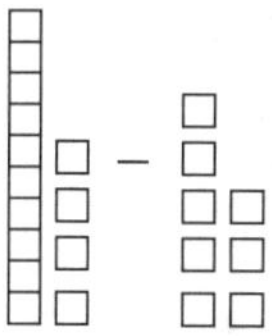 – =

Tens	Ones
–	

e. 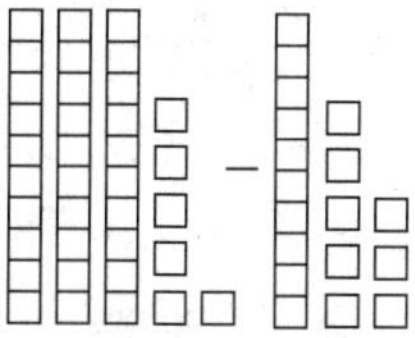 – =

Tens	Ones
–	

f. 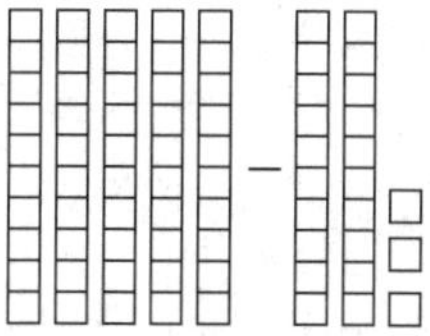 – =

Tens	Ones
–	

2. Regroup one ten for ten ones, then complete the problem. The first one has been done for you.

a. $\overset{1}{\cancel{2}}3 - 6 = 17$

b. 21 – 8

c. 23 – 16

d. 44 – 17

e. 52 – 19

f. 63 – 47

g. 70 – 45

h. 62 – 18

i. 47 – 19

j. 33 – 26

k. 52 – 18

l. 41 – 18

m. 31 – 18

n. 40 – 16

o. 51 – 15

3. a. There are 23 children in our class. 16 were sick today. How many children were there?

______ – ______ = ______

Tens	Ones
–	

b. James brought 65 cents to school. He spent 38 cents. How much was left?

______ – ______ = ______

Tens	Ones
–	

Name **Date**

1. **a.**

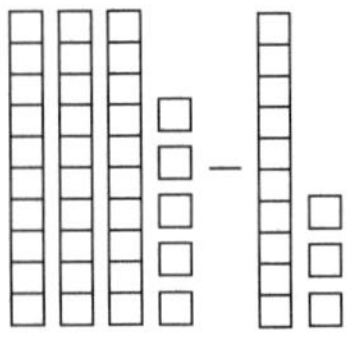

Tens ¦ Ones

b.

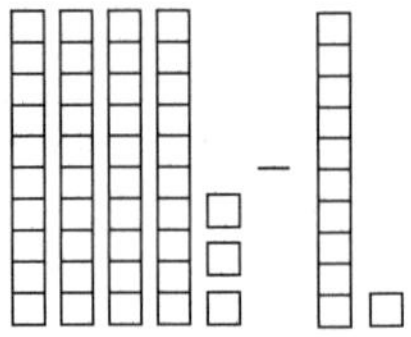

Tens ¦ Ones

2. Complete these subtraction wheels.

a.

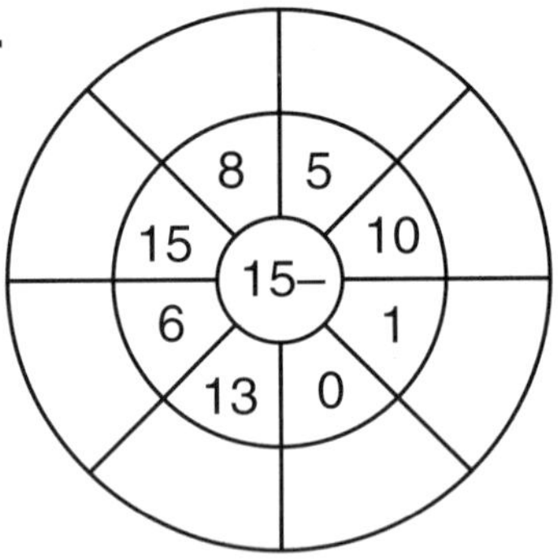

b.

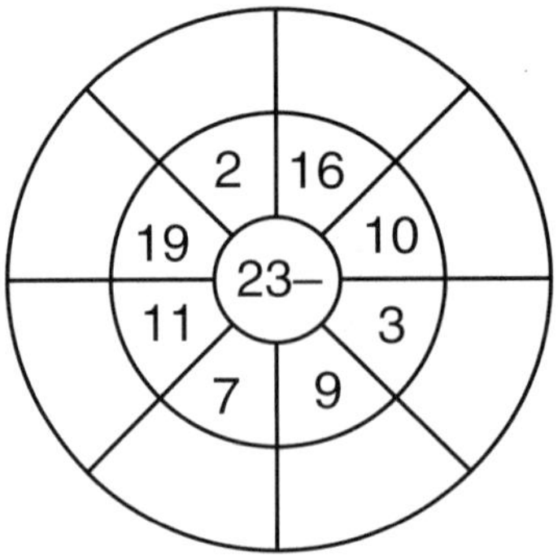

c.

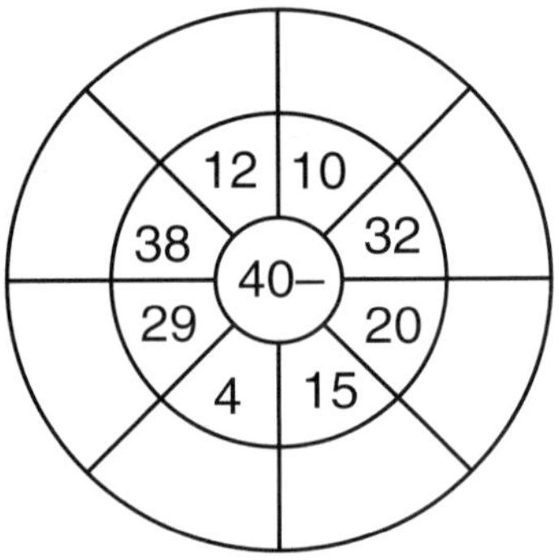

3. Complete these subtraction problems. Show all work.

a. 23 − 17

b. 30 − 18

c. 52 − 29

d. 61 − 48

e. 44 − 26

f. 73 − 34

g. 85 − 57

h. 96 − 37

4. Solve these subtraction problems.

a. Tim counted 73 purple cars, Carl counted 52. How many more did Tim count?

____

b. Bruce saved $92.00 during the week and Natalie saved $79.00. What was the difference in their savings?

____

c. Amanda read 54 books during the year. Paul only read 16. How many more books did Amanda read than Paul?

____

Name **Date**

Puzzle

Add extra lines to these subtraction words to make them true.

Example: THIRTY take away TEN equals TWENTY

1. **ΓIΓ I Y** minus **I IIRLL** equals **ΓORI V–SI \ LN**

2. **I Ⱶ IRI V–NINL** minus **NIIIΓ** equals **I HIRI V**

3. **I WLN I Y** minus **SI ** equals **I OURILΓII**

4. **SI \ I V–OIIL** take away **LIGII I** equals **ΓIΓ I Y–I IIRLΓ**

5. Write two of your own.

___

___

Game

Number of players: 2, 3, or 4

Aim: To be the first to reach zero

Equipment: 2 dice and scrap paper for each group

To Play: Each member of the group starts with 100 points. The players, in turn, roll the dice. The total of the two numbers are subtracted from his/her score. If a double is rolled (2, 2 or 5, 5 etc.) the turn is forfeited (score stays the same). If double one is rolled the player must start again from 100. The first player to reach zero is the winner.

Variation: The game can be played with 1, 2, or 3 dice depending on ability levels.

MORE SUBTRACTION PRACTICE

Unit 2

Regrouping
Problem solving
Calculations
Money
Estimating

Objectives

- *carry out operations on money*
- *use one or more strategies to solve mathematical problems*
- *generate and describe number patterns using a variety of strategies*
- *identify ones, tens, and hundreds*
- *use a calculator to carry out calculations*
- *check whether answers to problems are correct and sensible*
- *recognize the difference between estimate and exact amount*

Language

ones, tens, hundreds, working out, columns, subtraction, take away, equals, less than, greater than, estimate, exact, regrouping, number sentence, expensive, cheapest, rounding off

Materials/Resources

unit cubes (Base 10), blank paper, coins (optional), calculators, dice, colored pencils

Contents of Student Pages

** Materials needed for each reproducible student page*

Remember

Before starting ensure each student:

- ❑ *is aware of the importance of showing all work.*
- ❑ *understands the importance of working neatly to eliminate careless mistakes.*
- ❑ *realizes the importance of checking all answers.*

Additional Activities

- ❑ *Have students make their own number cross puzzles.*
- ❑ *Find different prices for goods in magazines. Compare prices by subtraction.*
- ❑ *Play a subtraction game for two or three players. Each player has a piece of paper and two dice. Start with 100 (200 etc.). In turns, roll the dice and subtract the total from the score. The first to get to zero is the winner.*

Answers

Page 32 Regrouping

1. a. 465 b. 388 c. 487 d. 587 e. 412 f. 205 g. 117 h. 184
2. a. < b. < c. > d. > e. > f. >
3. a.

−		
33	22	11
17	8	9
16	14	2

b.

−		
27	18	9
11	9	2
16	9	7

c.

−		
20	9	11
10	3	7
10	6	4

4. a.
```
  286
 − 39
  247
```
b.
```
  $120
 − $56
   $64
```

Page 33 Patterns

1. a. 66, 30, 37, 44, 23, 59, 84, 76, 39, 26, 45, 56
   b. 72, 36, 43, 50, 29, 65, 90, 82, 45, 32, 51, 62
   c. 55, 19, 26, 33, 12, 48, 73, 65, 28, 15, 34, 45
2. a. 91, 76, 70, 57, 39, 18
   b. 300, 244, 213, 168, 63, 1
3. a. 17, 27 b. 13, 26 c. 5, 10 d. 30, 40 e. 12, 18 f. 60, 80
4. a. 205 b. 399 c. 279 d. 43 e. 575 f. 219 g. 131 h. 414 i. 78

Page 34 Calculators

1. a.

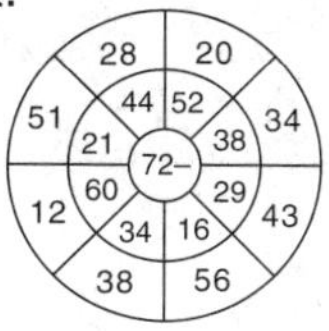

b.

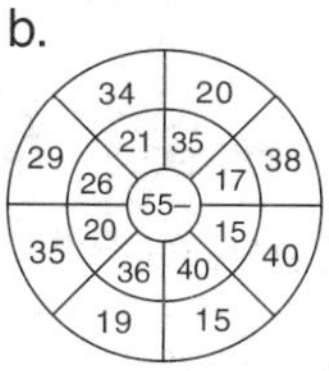

c.

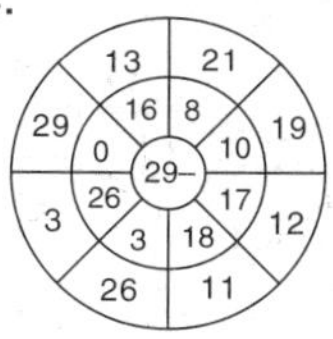

2. a. 4 b. 23 c. 9 d. 12 e. 56 f. 16

3. a. 53, 13
   b. 18, 40
   c. 173, 173, 114, 173, 59
   d. 153, 153, 21, 153, 132
   e. 221, 221, 16, 221, 205
   f. 105, 105, 88, 105, 17
4. a. 608, 610 b. 99, 100 c. 204, 200 d. 167, 170 e. 131, 130 f. 167, 170

Page 35 Money

1. a.
```
   70¢
 − 15¢
   55¢
```
b.
```
  100¢
 − 25¢
   75¢
```
c.
```
   60¢
 − 45¢
   15¢
```
d.
```
   80¢
 − 30¢
   50¢
```
2. a. $5.45 b. $1.15 c. $1.55
3. a. $1.15 b. $0.55 c. $1.05 d. $5.85 e. $7.77 f. $3.05 g. $6.35 h. $2.01

Page 36 Estimating

Make sure all estimates are reasonable. Below are the exact answers.

1. a. $181
   b. $77
   c. $97
   d. The microwave should be colored.
   e. The saucepan should have a cross on it.
2.

^{1}1	0	■	^{2}6	1	■	^{3}7	9	■	^{4}7
1	■	^{5}4	7	■	^{6}5	6	■	^{7}6	4
■	^{8}2	3	■	^{9}8	3	■	^{10}3	5	■
^{11}4	4	■	^{12}7	2	■	^{13}7	8	■	■

Page 37 Assessment

1. a. 26 b. 23 c. 15 d. 14 e. 12 f. 84 g. 31 h. 67 i. 305 j. 69
2. 133, 124, 88, 10, 0
3. a.
```
  $250
− $179
   $71
```
b.
```
  $156
 − $84
   $72
```
4.

−		
73	36	37
27	18	9
46	18	28

5.

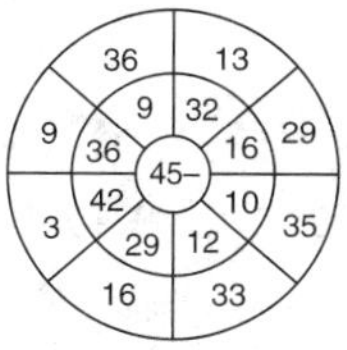

Name **Date**

1. Complete these problems.

a. 532 − 67	**b.** 444 − 56	**c.** 510 − 23	**d.** 672 − 85
e. 534 − 122	**f.** 320 − 115	**g.** 283 − 166	**h.** 404 − 220

2. Write < or > to make each statement true (< means less than, > means greater than).

a. 50 − 32 ☐ 40 − 4 **b.** 63 − 38 ☐ 58 − 16

c. 19 − 3 ☐ 22 − 10 **d.** 38 − 15 ☐ 40 − 22

e. 25 − 4 ☐ 33 − 16 **f.** 68 − 18 ☐ 49 − 4

3. Complete these subtraction squares. Subtract rows and columns.

a. (−)

33		11
17	8	
16		2

b. (−)

27		9
11	9	
	9	7

c. (−)

20	9	
10		7
	6	4

4. Show all your work for these problems.

a. There are 286 children at the local school. Thirty-nine children are away with chicken pox. How many children are at school?

	Hundreds	Tens	Ones
−			

b. Chloe's mom spent $56 at the shop. If she had $120 before she left home, how much change did she have left in her purse?

	Hundreds	Tens	Ones
−			

Name	**Date**

1. Complete the table. Regroup when necessary.

	–	82	46	53	60	39	75	100	92	55	42	61	72
a.	16					**23**							
b.	10												
c.	27												

2. Follow the arrows to find your answer.

a. (100) — −9 → ( ) — −15 → ( ) — −6 → ( ) — −13 → ( ) — −18 → ( ) — −21 → ( )

b. (400) — −100 → ( ) — −56 → ( ) — −31 → ( ) — −45 → ( ) — −105 → ( ) — −62 → ( )

3. Write the first two numbers of each number pattern.

a. _____ , _____ , 37, 47, 57

b. _____ , _____ , 39, 52, 65

c. _____ , _____ , 15, 20, 25

d. _____ , _____ , 50, 60, 70

e. _____ , _____ , 24, 30, 36

f. _____ , _____ , 100, 120, 140

4. Be sure to regroup carefully.

a. $\begin{array}{r} 630 \\ -\ 425 \\ \hline \end{array}$

b. $\begin{array}{r} 789 \\ -\ 390 \\ \hline \end{array}$

c. $\begin{array}{r} 600 \\ -\ 321 \\ \hline \end{array}$

d. $\begin{array}{r} 732 \\ -\ 689 \\ \hline \end{array}$

e. $\begin{array}{r} 901 \\ -\ 326 \\ \hline \end{array}$

f. $\begin{array}{r} 409 \\ -\ 190 \\ \hline \end{array}$

g. $\begin{array}{r} 430 \\ -\ 299 \\ \hline \end{array}$

h. $\begin{array}{r} 803 \\ -\ 389 \\ \hline \end{array}$

i. $\begin{array}{r} 216 \\ -\ 138 \\ \hline \end{array}$

Name **Date**

1. Complete the number wheels.

a.

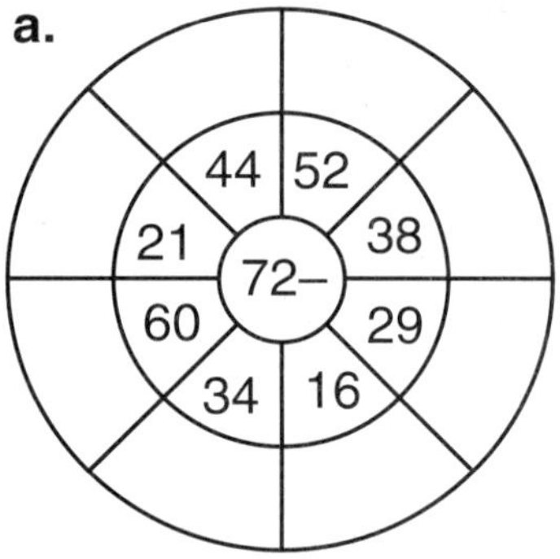

b.

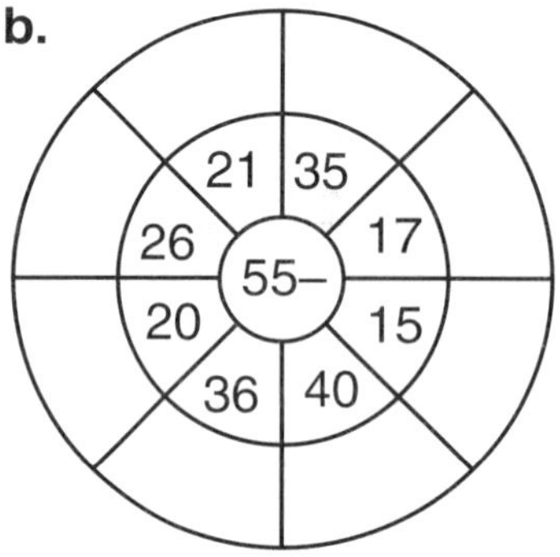

c.

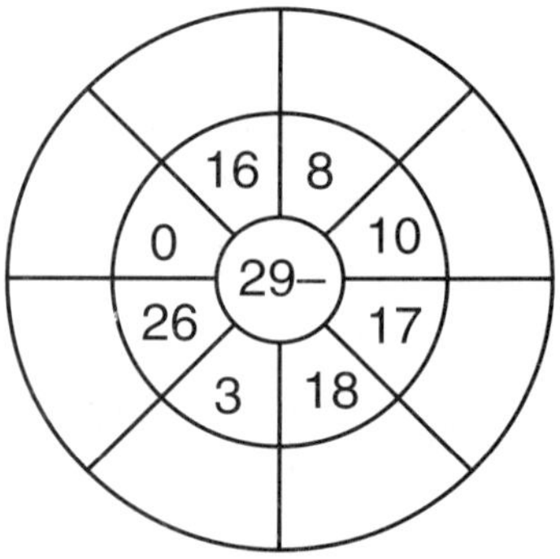

2. Use your calculator. Find how many times each number can be subtracted.

a. 7 subtracted from 28 ______ **b.** 5 subtracted from 115 ______

c. 13 subtracted from 117 ______ **d.** 27 subtracted from 324 ______

e. 3 subtracted from 168 ______ **f.** 30 subtracted from 480 ______

3. Write two more number sentences that relate to the first one.

a. 53 + 13 = 66	66 – 13 = ______	66 – 53 = ______
b. 40 + 18 = 58	58 – 40 = ______	58 – 18 = ______
c. 114 + 59 = ______	______ – 59 = ______	______ – 114 = ______
d. 132 + 21 = ______	______ – 132 = ______	______ – 21 = ______
e. 205 + 16 = ______	______ – 205 = ______	______ – 16 = ______
f. 17 + 88 = ______	______ – 17 = ______	______ – 88 = ______

4. Work out these answers and then round them to the nearest ten.

a. 830 − 222 Round

b. 798 − 699 Round

c. 502 − 298 Round

d. 666 − 499 Round

e. 532 − 401 Round

f. 365 − 198 Round

Name **Date**

1. Write a number sentence to find the answer.

a. 

–

b. 

–

c. 

–

d. 

–

2. Solve these money problems. Show all work.

a. Trish went to the bank to withdraw some money from her savings account. She had $10.00 in her account and withdrew $4.55. How much was her new balance?

____

b. Mick had $5.00 in his wallet. He lent $3.85 to George. How much did he still have in his wallet?

____

c. Stacey went to the ice-cream shop and bought an ice cream for $1.95. What was her change from $3.50?

____

3. Work out the problems. Be sure to check for mistakes.

a. $2.55 – $1.40

b. $1.80 – $1.25

c. $5.05 – $4.00

d. $10.50 – $4.65

e. $11.45 – $3.68

f. $15.95 – $12.90

g. $9.00 – $2.65

h. $18.00 – $15.99

Name **Date**

1. Estimate the difference in prices. Then find the exact difference. Show your work.

lamp
$62

clock
$123

bowl
$54

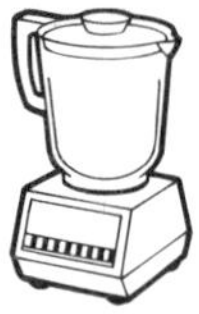
blender
$46

saucepan
$39

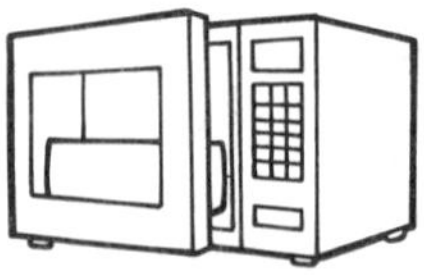
microwave
$220

a. What is the difference in price between the saucepan and microwave?

Estimate = ______ Exact answer = ______

b. What is the difference in price between the clock and the blender?

Estimate = ______ Exact answer = ______

c. What is the difference in price between the microwave and clock?

Estimate = ______ Exact answer = ______

d. Color the most expensive item.

e. Cross out the cheapest item.

2. Complete the subtraction cross number puzzle.

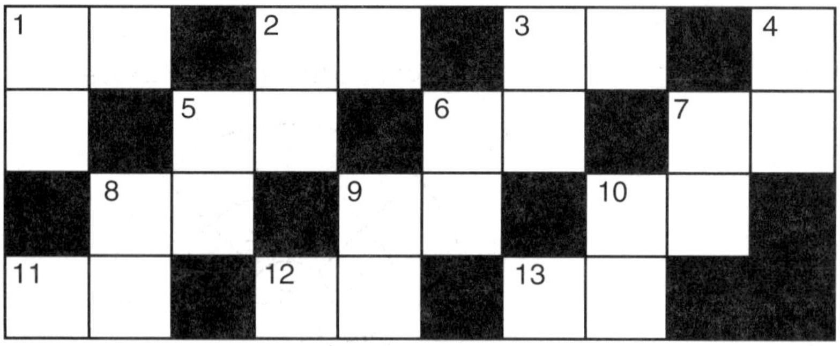

Across

1. 100 – 90
2. 100 – 39
3. 100 – 21
5. 100 – 53
6. 100 – 44
7. 100 – 36
8. 100 – 77
9. 100 – 17
10. 100 – 65
11. 100 – 56
12. 100 – 28
13. 100 – 22

Down

1. 100 – 89
2. 100 – 33
3. 100 – 24
4. 100 – 26
5. 100 – 57
6. 100 – 47
7. 100 – 35
8. 100 – 76
9. 100 – 18
10. 100 – 62

Name **Date**

1. Show all work.

a. 53 − 27	**b.** 62 − 39	**c.** 45 − 30	**d.** 63 − 49	**e.** 75 − 63
f. 253 − 169	**g.** 300 − 269	**h.** 401 − 334	**i.** 604 − 299	**j.** 724 − 655

2. Follow the arrows to the final answer.

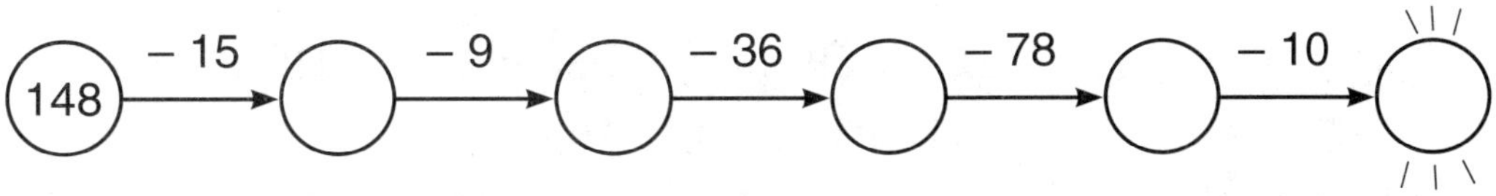

3. Read the problems carefully. Show all work in the space provided.

a. Simon bought a new blue bicycle for Lucy for $250. Bruce bought a new blue bicycle for Charlie for $179. What was the difference in price of the bicycle?

b. Patty found a new pair of shoes on sale for $84. They were originally selling for $156. How much did Patty save?

4. Complete the subtraction square.

–			
	73	36	
	27	18	

5. Complete the number wheel.

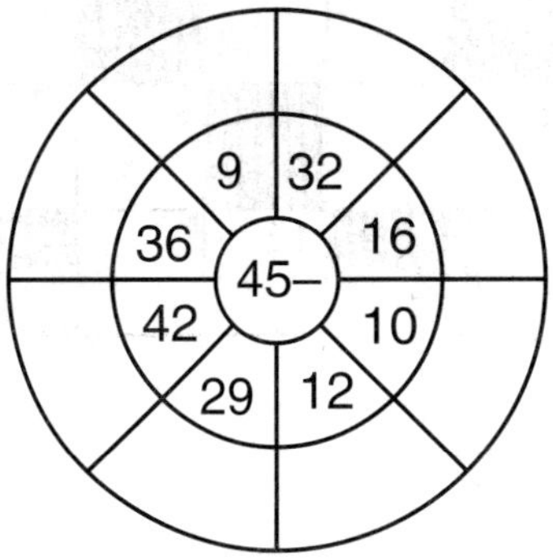

MULTIPLICATION

Multiplication is explored through factors, multiples, products, grouping, and number patterns. Problem solving and both vertical and horizontal algorithms add further practice. Algorithms extend to three-digit numbers multiplied by one-digit numbers, and two-digit numbers by two-digit numbers (with and without regrouping). Table facts are reinforced through money problems, and completing grids. The correct mathematical language for multiplication is encouraged. One activity page contains the instructions for a multiplication game for 2–4 players and a playing board. There is also an activity that reinforces the multiples of 8. Two assessment pages are also included.

MULTIPLICATION PRACTICE

Unit 1

Problem solving
Products
Factors
Parentheses

Objectives

- recognize patterns in multiplication tables
- use a variety of strategies to calculate and represent solutions to multiplication problems
- use written methods to multiply whole numbers by single digit numbers and multiples of ten
- check whether answers to problems are correct and sensible
- understand the term "factors" and apply this understanding
- recall multiplication facts up to 10 x 10
- write complete equivalent number statements using the = sign
- is able to show all work

Language

ones, tens, hundreds, times, multiplication, multiply, equals, groups, factors, tables, parentheses, product

Materials/Resources

counters, 1–10 spinners, polyhedral dice, colored pencils

Contents of Student Pages

* *Materials needed for each reproducible page*

Remember

Before starting ensure each student:

- ❑ *realizes the importance of knowing the multiplication tables well.*
- ❑ *reads problem solving questions twice.*
- ❑ *checks answers.*
- ❑ *understands that neatness and care with working out the problem is very important.*

Additional Activities

- ❑ *Make sets of multiplication flash cards. Two players test each other to see who gets the most correct answers.*
- ❑ *Play "Speed Test." The teacher calls out multiplication questions and students write the answers. Students must keep up with the teacher to score 20 out of 20.*
- ❑ *In small groups, make two dice numbered 4–9. Each player throws the dice and finds the product of the two numbers rolled. The highest scoring person wins.*
- ❑ *A clock face can be used to help count by fives.*

Answers

Page 41 Problem Solving

1. a. 9 x 7 = 63 hours
   b. 3 x 4 = 12 hours
   c. 8 x 9 = 72 flowers
   d. 7 x 8 = 56 oranges
2. a. 6 x 10 = 60¢
   b. 9 x 5 = 45¢
   c. 25 x 4 = 100¢ or $1.00
   d. 7 x 1 = $7.00
   e. 5 x 1 = 5¢
3. b. 6 x 8 = 8 x 6 = 48
   c. 6 x 5 = 5 x 6 = 30
   d. 7 x 3 = 3 x 7 = 21
   e. 5 x 4 = 4 x 5 = 20
   f. 9 x 2 = 2 x 9 = 18
   g. 7 x 6 = 6 x 7 = 42
   h. 6 x 4 = 4 x 6 = 24
   i. 8 x 4 = 4 x 8 = 32
   j. 8 x 7 = 7 x 8 = 56
4. a. 20, 36, 28, 8, 16, 12, 32, 4, 24
   b. 35, 5, 25, 30, 40, 15, 45, 10, 20

Page 42 Products

1. a. 12 b. 48 c. 12 d. 25 e. 20 f. 35 g. 72 h. 36 i. 42 j. 16 k. 36 l. 10
2. a. 49 b. 36 c. 16 d. 21 e. 35 f. 48 g. 56 h. 70
   i. Balloons b, c, f, g, and h should be colored red.
   j. Balloons a, d, and e should be colored blue.
3. a. 84 b. 45 c. 92 d. 108 e. 55 f. 60 g. 57 h. 63 i. 90 j. 119
4. a. 42, 24, 12, 30, 18, 48, 36, 54, 60
   b. 70, 21, 28, 14, 35, 49, 63, 42, 56

Page 43 Factors

1. a. 4 b. 6 c. 5 d. 7 e. 6 f. 9 g. 12 h. 4 i. 3
2. b. 1 x 24, 2 x 12, 3 x 8, 4 x 6
   c. 1 x 36, 2 x 18, 3 x 12, 4 x 9, 6 x 6
   d. 1 x 12, 2 x 6, 3 x 4
   e. 1 x 10, 2 x 5
   f. 1 x 30, 2 x 15, 3 x 10, 5 x 6
   g. 1 x 20, 2 x 10, 4 x 5
3. a. 5, 10, 15, 20, 25, 30, 35, 40, 45, 50
   b. 7, 14, 21, 28, 35, 42, 49, 56, 63, 70
4. a. 40, 56, 24, 48, 16, 80, 32, 72, 64
   b. 90, 72, 36, 81, 63, 54, 18, 45, 27

Page 44 More Problem Solving

1. a. 14 x 5 = 70 — 5 x 14 = 70 miles
   b. 12 x 4 = 48 — 4 x 12 = 48 legs
   c. 19 x 9 = 171 — 19 x 9 = 171 pencils
   d. 15¢ x 8 = 120¢ or $1.20 — 15¢ x 8 = 120¢ or $1.20
2. a. 72 b. 112 c. 96 d. 135 e. 171 f. 132 g. 168 h. 87 i. 119 j. 156
3. a. 40, 50, 90, 80, 20, 70, 30, 60, 100
   b. 0, 0, 0, 0, 0, 0, 0, 0, 0

Page 45 Multiply It!

1. a. 19 b. 37 c. 51 d. 28 e. 50 f. 20 g. 90 h. 29 i. 31 j. 19 k. 12 l. 84
2. a. 10 b. 7 c. 8 d. 7 e. 5 f. 9 g. 6 h. 2 i. 0
3. a. 10 b. 12, 21 c. 30, 60 d. 35, 42 e. 30, 35, 40 f. 32, 48, 56
4. a. 460 b. 812 c. 944 d. 896 e. 999 f. 770 g. 904 h. 570 i. 464 j. 357 k. 600 l. 492 m. 726 n. 426 o. 604

Page 46 Assessment

1. a. 45 b. 42 c. 147 d. 78 e. 176 f. 777 g. 189 h. 140 i. 810 j. 459
2. a. 1, 24, 2, 12, 3, 8, 4, 6
   b. 1, 8, 2, 4
   c. 1, 30, 2, 15, 3, 10, 5, 6
   d. 1, 45, 3, 15, 5, 9
   e. 1, 16, 2, 8, 4
   f. 1, 19
3. a. 45¢ x 7 = 315¢ or $3.15 — 45¢ x 7 = 315¢ or $3.15
   b. 8 x 9 = 72
      8 x 9 = 72 trees
4. a. 6 x 8 = 8 x 6 = 48
   b. 9 x 3 = 3 x 9 = 27
   c. 2 x 9 = 9 x 2 = 18
   d. 5 x 6 = 6 x 5 = 30
   e. 7 x 8 = 8 x 7 = 56
   f. 4 x 5 = 5 x 4 = 20
5. a. 35, 63, 42, 0, 14, 21, 56, 7
   b. 12, 36, 24, 20, 0, 28, 32, 16
6. a. 33 b. 40 c. 108 d. 13 e. 31 f. 30 g. 50 h. 55 i. 82

Page 47 Activity—Joining 4/Color the Path

8, 72, 112, 16, 64, 72, 104, 24, 48, 56, 88, 80, 32, 40, 120, 96, 128

Name **Date**

1. Solve the multiplication problems by writing them out as number sentences.

a. The man in the fruit shop works 9 hours a day. He works 7 days a week. How many hours does he work in one week?

_______ x _______ = _______

c. The florist made 8 bunches of flowers. Each bunch had 9 flowers tied together. How many flowers did she use altogether?

_______ x _______ = _______

b. Mike rides his bike 3 hours each weekend. How many hours does he ride in one month?

_______ x _______ = _______

d. There are 7 oranges in a bag. How many oranges in 8 bags?

_______ x _______ = _______

2. What is the value of each?

a.

6 coins

___ x ___ = ___

b.

9 coins

___ x ___ = ___

c.

4 coins

___ x ___ = ___

d.

7 coins

___ x ___ = ___

e.

5 coins

___ x ___ = ___

3. Complete the number sentences by reversing them. The first one has been done for you.

a. 4 x 3 = 3 x 4 = 12

c. 6 x 5 = ___ x ___ = ___

e. 5 x 4 = ___ x ___ = ___

g. 7 x 6 = ___ x ___ = ___

i. 8 x ___ = 4 x ___ = ___

b. 6 x ___ = 8 x ___ = ___

d. 7 x 3 = ___ x ___ = ___

f. 9 x 2 = ___ x ___ = ___

h. ___ x 4 = ___ x 6 = ___

j. 8 x 7 = ___ x ___ = ___

4. **a.**

x	5	9	7	2	4	3	8	1	6
4									

b.

x	7	1	5	6	8	3	9	2	4
5									

Name **Date**

1. Find the products of these numbers.

a. 3 and 4 = ____ **b.** 6 groups of 8 = ____ **c.** 6 times 2 = ____ **d.** 5 and 5 = ____

e. 5 and 4 = ____ **f.** 7 groups of 5 = ____ **g.** 8 groups of 9 = ____ **h.** 6 times 6 = ____

i. 6 and 7 = ____ **j.** 4 x 4 = ____ **k.** 4 x 9 = ____ **l.** 2 groups of 5 = ____

2.

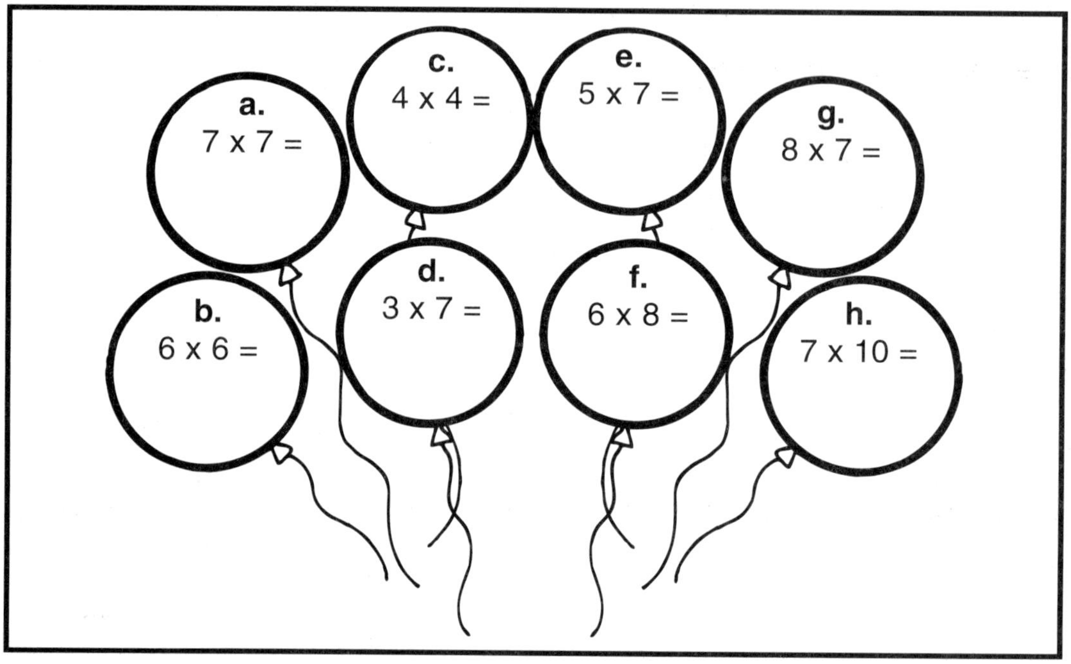

i. Color the even numbered balloons red.

j. Color the odd numbered balloons blue.

3.

a. 12 x 7 **b.** 15 x 3 **c.** 23 x 4 **d.** 18 x 6 **e.** 11 x 5

f. 10 x 6 **g.** 19 x 3 **h.** 9 x 7 **i.** 15 x 6 **j.** 17 x 7

4. **a.**

x	7	4	2	5	3	8	6	9	10
6									

b.

x	10	3	4	2	5	7	9	6	8
7									

Name **Date**

1. Find the factors that are missing.

a. 6 x ____ = 24 **b.** ____ x 5 = 30 **c.** 4 x ____ = 20

d. 5 x ____ = 35 **e.** ____ x 7 = 42 **f.** 8 x ____ = 72

g. 2 x ____ = 24 **h.** 3 x ____ = 12 **i.** ____ x 7 = 21

2. Find the factors of these numbers. The first one had been done for you.

a. 15 = 1 x 15, 5 x 3

b. 24 = ____ x ____, ____ x ____, ____ x ____, ____ x ____

c. 36 = ____ x ____, ____ x ____, ____ x ____, ____ x ____, ____ x ____

d. 12 = ____ x ____, ____ x ____, ____ x ____

e. 10 = ____ x ____, ____ x ____

f. 30 = ____ x ____, ____ x ____, ____ x ____, ____ x ____

g. 20 = ____ x ____, ____ x ____, ____ x ____

3. Follow the arrows and add the number in the first box to each box on the path.

a. Start

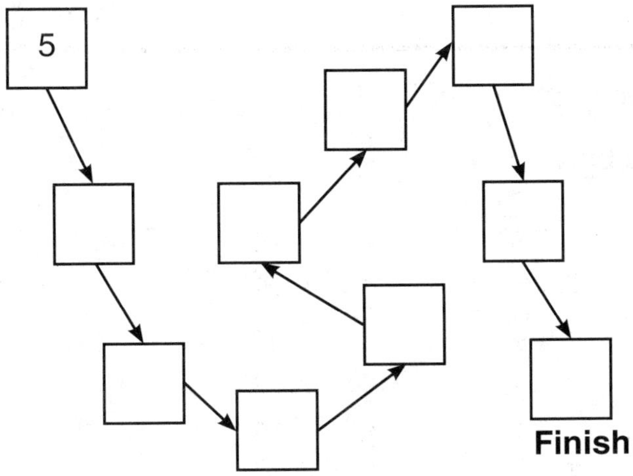

b. Start

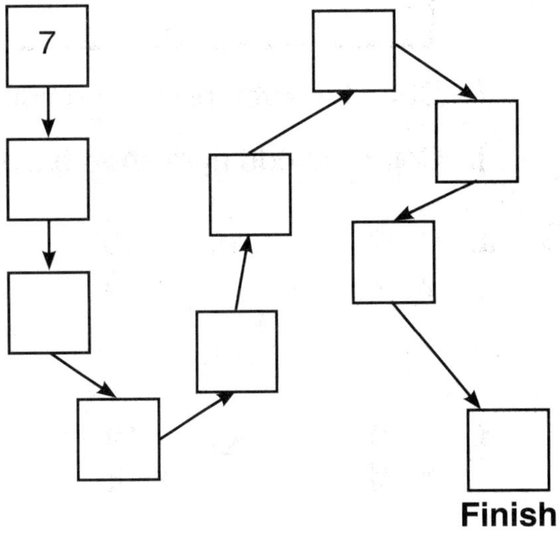

4. a.

x	5	7	3	6	2	10	4	9	8
8									

b.

x	10	8	4	9	7	6	2	5	3
9									

Name **Date**

1. Solve these problems.

a. Mel walks 5 miles a day. How far does she walk in 14 days?

x

___ x ___ = ___

b. A dog has 4 legs. How many legs do a dozen dogs have?

x

___ x ___ = ___

c. There were 19 children in the class. Each has 9 pencils in their pencil case. How many pencils were there altogether?

x

___ x ___ = ___

d. Apples cost 15 cents each. How much will it cost to buy 8 apples?

x

___ x ___ = ___

2.
a. 12 x 6
b. 16 x 7
c. 12 x 8
d. 15 x 9
e. 19 x 9
f. 22 x 6
g. 24 x 7
h. 29 x 3
i. 17 x 7
j. 26 x 6

3. **a.**

x	4	5	9	8	2	7	3	6	10
10									

b.

x	1	5	4	7	3	6	9	2	8
0									

Name **Date**

1. Solve these problems. Remember to work out the parentheses first.

a. 7 + (3 x 4) = ____ **b.** (6 x 5) + 7 = ____ **c.** (6 x 9) – 3 = ____

d. (8 x 2) + (3 x 4) = ____ **e.** (2 x 10) + (3 x 10) = ____ **f.** (5 x 2) + (5 x 2) = ____

g. (9 x 9) + 9 = ____ **h.** (8 x 3) + 5 = ____ **i.** (7 x 3) + 10 = ____

j. 10 + (3 x 3) = ____ **k.** (3 x 3) + 3 = ____ **l.** (6 x 7) + (6 x 7) = ____

2. Find the missing number facts.

a. 6 x ____ = 60 **b.** 7 x ____ = 49 **c.** ____ x 8 = 64

d. 10 x ____ = 70 **e.** 5 x ____ = 25 **f.** 9 x ____ = 81

g. 4 x ____ = 24 **h.** 7 x ____ = 14 **i.** 10 x ____ = 0

3. Fill in the spaces to complete the patterns.

a. 2, 4, 6, 8, ____ , 12 **b.** 6, 9, ____, 15, 18, ____

c. 20, ____ , 40, 50, ____ , 70 **d.** 21, 28, ____ , ____ , 49, 56

e. 20, 25, ____ , ____ , ____ **f.** 16, 24, ____ , 40, ____ , ____

4.

a. 115 x 4	**b.** 116 x 7	**c.** 118 x 8	**d.** 112 x 8	**e.** 111 x 9
f. 110 x 7	**g.** 113 x 8	**h.** 114 x 5	**i.** 116 x 4	**j.** 119 x 3
k. 200 x 3	**l.** 123 x 4	**m.** 121 x 6	**n.** 142 x 3	**o.** 151 x 4

Name **Date**

1.

a. 9 x 5	**b.** 6 x 7	**c.** 21 x 7	**d.** 26 x 3	**e.** 44 x 4
f. 111 x 7	**g.** 21 x 9	**h.** 35 x 4	**i.** 135 x 6	**j.** 153 x 3

2. List the factors for the following numbers.

a. 24 ____________ **b.** 8 ____________

c. 30 ____________ **d.** 45 ____________

e. 16 ____________ **f.** 19 ____________

3. Solve these problems and show your work.

a. James bought 7 postcards on his holiday. Each postcard cost 45 cents. How much did they cost altogether?

x ___ x ___ = ___

b. Farmer Jo had 8 rows of avocado trees. In each row there were 9 trees. How many avocado trees were there in the orchard?

x ___ x ___ = ___

4. Complete the number sentences by reversing them.

a. 6 x 8 = 8 x ___ = 48

b. 9 x 3 = ___ x ___ = ___

c. 2 x 9 = ___ x ___ = ___

d. 5 x ___ = 6 x ___ = ___

e. 7 x 8 = ___ x ___ = ___

f. 4 x 5 = ___ x ___ = ___

5. **a.**

x	5	9	6	0	2	3	8	1
7								

b.

x	3	9	6	5	0	7	8	4
4								

6. **a.** 8 + (5 x 5) = _____ **b.** (6 x 5) + 10 = _____ **c.** (10 x 10) + 8 = _____

d. 2 + 2 + (3 x 3) = _____ **e.** 9 + (6 x 3) + 4 = _____ **f.** (7 x 4) + 2 = _____

g. 4 + 4 + (6 x 7) = _____ **h.** 5 + (5 x 10) = _____ **i.** (9 x 9) + 1 = _____

Name **Date**

Joining 4 (Game)

Number of Players: 2 or 2 teams of 2 (4)

Equipment: Two 1–10 spinners or polyhedral dice, plastic counters (different colors for each team), pencils, the grid below enlarged

Aim: To be the first to have four counters next to each other in a row, column, or diagonal

How to Play: Roll the dice or spinners and multiply the two numbers. Cover this number with a counter.

Strategies: Try to block your opponent. If the number you have spun is covered by your opponent and there is not another available, you may remove your opponent's counter and put yours down.

1	49	40	8	36	14	64	3
24	48	12	80	30	36	10	90
81	32	24	21	64	45	36	27
9	24	63	2	72	15	16	18
24	28	30	54	54	7	60	36
20	56	16	40	50	30	63	4
5	100	49	9	25	18	56	63
42	36	6	70	12	35	100	81

Color the Path (Activity)

Color the path to show multiples of 8. They do not appear in order. Start at 8 and finish at 128.

8	13	64	72	104	24	48	49
72	112	16	52	54	55	56	65
6	81	2	11	90	80	88	99
12	201	87	57	17	32	93	19
101	131	128	96	120	40	60	70

MORE MULTIPLICATION PRACTICE

Unit 2

Grouping
Problem solving
Factors
Multiplying by 10, 100, and 1,000

Objectives

- recall multiplication facts
- is able to show all work
- recognize patterns in multiplication tables
- use a variety of strategies to calculate and represent solutions to multiplication problems
- check whether answers to problems are correct and sensible
- demonstrate an understanding of factors
- use written methods of multiplication to solve problems relating to money

Language

ones, tens, hundreds, times, groups, multiples, multiplication, factors, equals, tables, product, true, false, parentheses

Materials/Resources

colored pencils, calculators

Contents of Student Pages

* Materials needed for each reproducible student page

Remember

Before starting ensure each student:

- ❑ knows times table multiples well.
- ❑ checks answers.
- ❑ reads problem solving question twice.
- ❑ understands that neatness and working the problem out can affect final answers.

Additional Activity

- *Allow the students to write their own test papers for each other. Encourage them to mark each others work. This develops their checking skills.*

Answers

Page 50 Groups

1. a. 6 groups of 6 — $6 \times 6 = 36$
   b. 7 groups of 8 — $7 \times 8 = 56$
   c. 9 groups of 2 — $9 \times 2 = 18$
   d. 4 groups of 10 — $4 \times 10 = 40$

2. a.
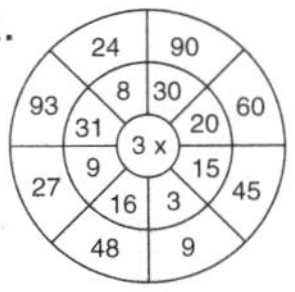

b.
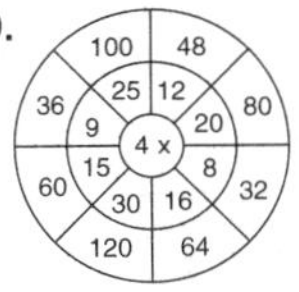

c.
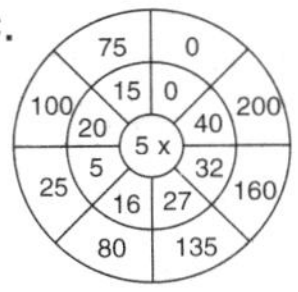

3. a. 1,230 d. 1,848 g. 1,668 j. 1,192 m. 1,270
   b. 438 e. 1,098 h. 875 k. 3,180 n. 344
   c. 702 f. 2,842 i. 2,336 l. 3,292 o. 792

Page 51 Problem Solving

1. a. (5 x 12) + (6 x 10) = $1.20
   b. (7 x 8) + (6 x 9) = 110
   c. (4 x 9) + (9 x 12) = 144
   d. (7 x 6) + (10 x 2) = $62
2. a. 8 x 12 = 12 x 8 = 96
   b. 9 x 8 = 8 x 9 = 72
   c. 10 x 11 = 11 x 10 = 110
   d. 12 x 10 = 10 x 12 = 120
   e. 8 x 20 = 20 x 8 = 160
   f. 14 x 10 = 10 x 14 = 140
   g. 7 x 13 = 13 x 7 = 91
   h. 6 x 20 = 20 x 6 = 120
3. a. 13, 26, 39, 52, 65, 78, 91, 104, 117, 130, 143, 156
   b. 20, 40, 60, 80, 100, 120, 140, 160, 180, 200, 220, 240
4. a. 54 e. 105 i. 44
   b. 130 f. 54 j. 85
   c. 99 g. 300 k. 63
   d. 84 h. 96 l. 120

Page 52 Factors and Multiples

1. a. 1, 15, 3, 5
   b. 1, 36, 2, 18, 3, 12, 4, 9, 6
   c. 1, 100, 2, 50, 4, 25, 10, 5, 20
   d. 1, 20, 2, 10, 4, 5
   e. 1, 64, 2, 32, 4, 16, 8
   f. 1, 19
2. a. 16, 64, 32, 8
   b. 24, 36, 60, 48
   c. 30, 100, 60, 70
3. a. $9.18 e. $44.24 i. $57.12
   b. $13.75 f. $35.28 j. $37.74
   c. $31.72 g. $23.52 k. $9.15
   d. $72.00 h. $15.18 l. $9.24
4. a. 114 e. 61 i. 128
   b. 96 f. 80 j. 33
   c. 136 g. 50 k. 188
   d. 67 h. 49 l. 120

Page 53 Multiply It!

1. a. 30 d. 230 g. 1,600 j. 3,000 m. 5,310
   b. 80 e. 330 h. 4,900 k. 14,000 n. 26,300
   c. 100 f. 400 i. 5,600 l. 230,000 o. 72,000
2. a. 30 c. 19 e. 2 g. 16
   b. 7 d. 3 f. 6 h. 35
3. a. 10 x 10¢ = 100¢ or $1.00
   b. 22 x 5¢ = 110¢ or $1.10
   c. 14 x 20¢ = 280¢ or $2.80
   d. 19 x 10¢ = 190¢ or $1.90
   e. 14 x $1 = $14.00
4. a. 14, 700, 224, 56, 189, 70, 84, 105, 140
   b. 16, 800, 256, 64, 216, 80, 96, 120, 160
   c. 18, 900, 288, 72, 243, 90, 108, 135, 180
   d. 20, 1,000, 320, 80, 270, 100, 120, 150, 200
   e. 32, 1,600, 512, 128, 432, 160, 192, 240, 320

Page 54 Multiplication By Tens

1. a. 864 d. 621 g. 1,353 j. 374 m. 306
   b. 704 e. 1,008 h. 1,122 k. 1,188 n. 375
   c. 396 f. 2,530 i. 1,416 l. 726 o. 2,262
2. a. True d. False g. False
   b. False e. True h. False
   c. False f. False
3. a. 77 d. 463,000 f. 1,104
   b. 81 e. 17 x 5 = 85 or 5 x 17 = 85 g. 221
   c. 9

Page 55 More Multiplication By Tens

1. a. 20 d. 6 g. 40
   b. 7 e. 9 h. 100
   c. 10 f. 30 i. 20
2. a.
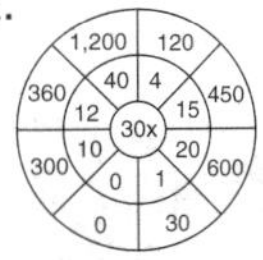

b.
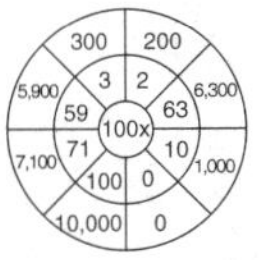

c.
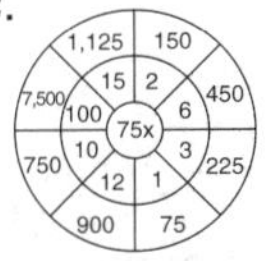

3. a. 1,458 c. 399 e. 5,427 g. 4,697 i. 2,508
   b. 270 d. 1,760 f. 2,784 h. 4,018 j. 3,658
4. 9; 36; 648; 4,536; 22,680; 22,680

Page 56 Assessment

1. a. 63 c. 944 e. 2,208 g. 774 i. 3,285
   b. 108 d. 1,930 f. 552 h. 1,155 j. 1,404
2. a. 1, 5, 25 e. 360
   b. 1, 100, 2, 50, 4, 25, 5, 20, 10 f. 1,020
   c. 230 g. 300
   d. 96 h. 73,010
3. a. (12 x $0.60) + (6 x $1.00) = $13.20
   b. (3 x 15¢) + (2 x 20¢) + (3 x 70¢) = $2.95
4. a.
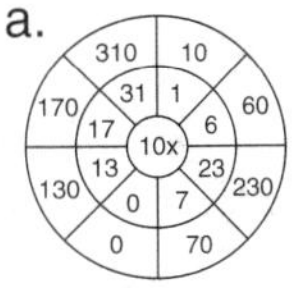

b.
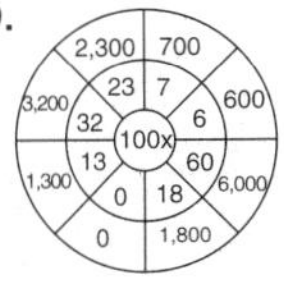

c.
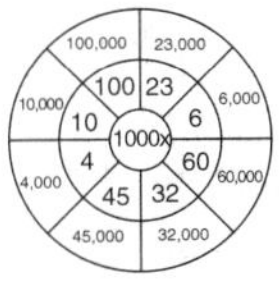

Name ______________________ **Date** ______________________

1. Complete.

a.

x

_____ groups of _____

b.

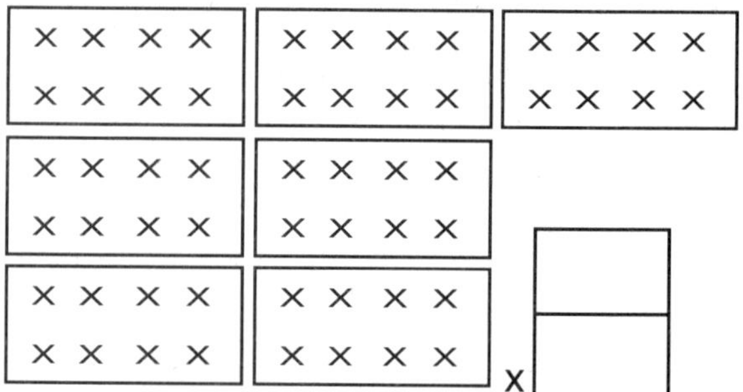

x

_____ groups of _____

c.

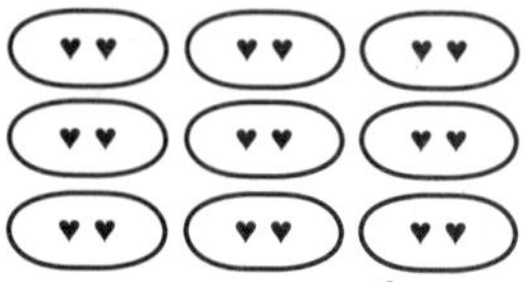

x

_____ groups of _____

d.

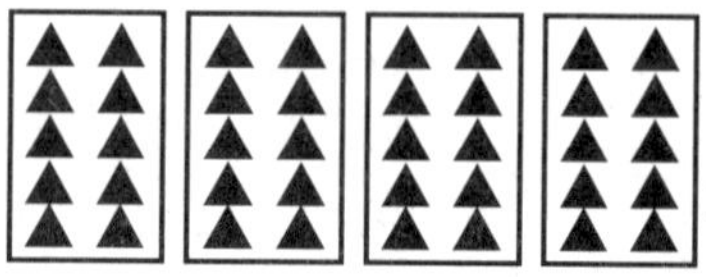

x

_____ groups of _____

2. Complete the number wheels.

a.

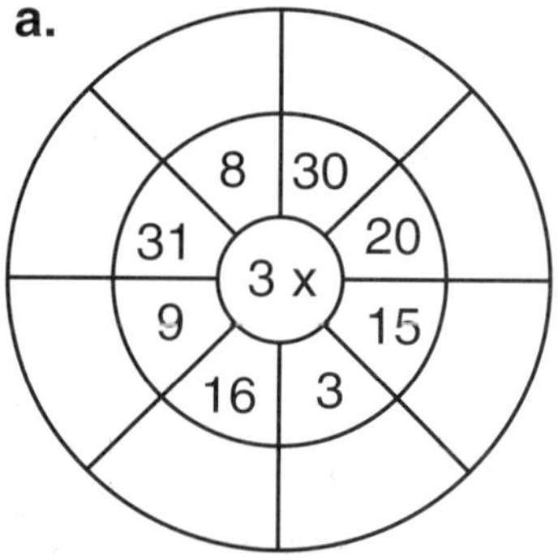

b.

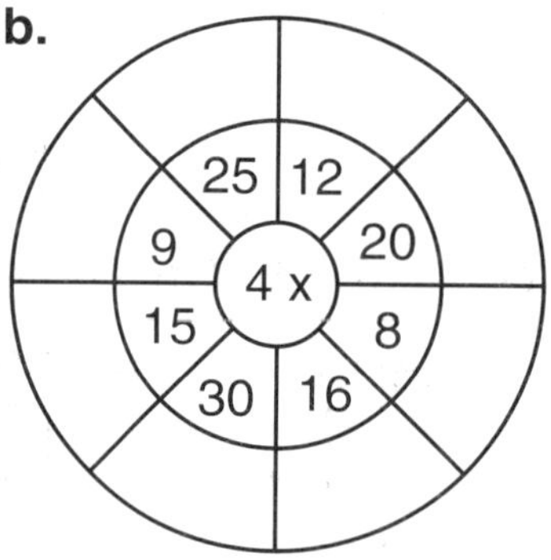

c.

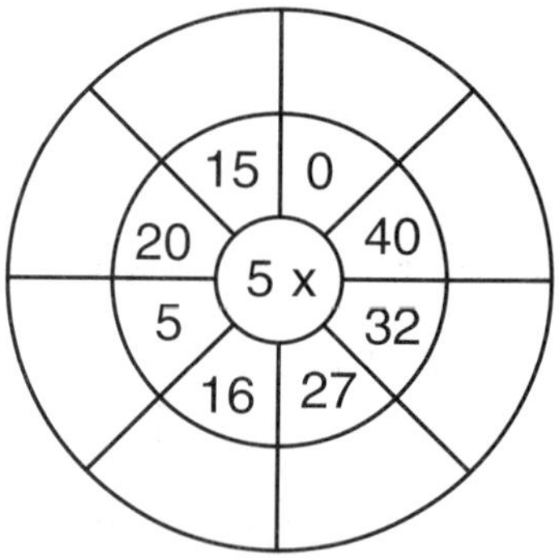

3. Watch the columns and show all work. The first one has been done for you.

a. 246 (carried: 2 3) x 5 = 1230

b. 146 x 3

c. 78 x 9

d. 264 x 7

e. 183 x 6

f. 406 x 7

g. 556 x 3

h. 175 x 5

i. 292 x 8

j. 149 x 8

k. 636 x 5

l. 823 x 4

m. 254 x 5

n. 172 x 2

o. 99 x 8

Name **Date**

1. Read the problems carefully. Use parentheses to help solve each problem.

a. I bought 5 apples at 12¢ each and 6 oranges at 10¢ each. How much did I spend altogether?

(___ x ___) + (___ x ___) = ___

b. In one paddock there were 7 rows of 8 mango trees. In another there were 6 rows of 9 apple trees. How many fruit trees altogether?

(___ x ___) + (___ x ___) = ___

c. Four children in third grade received 9 awards. In fourth grade 9 children received 12 awards. How many awards were given out altogether?

(___ x ___) + (___ x ___) = ___

d. Penny ordered 7 yards of fabric at $6 a yard and 10 yards of fabric at $2 a yard. How much did she spend on fabric?

(___ x ___) + (___ x ___) = ___

2. Reverse the number sentences so that the number facts are equal.

a.	8 x 12	=	___	x	___	=	___
b.	9 x 8	=	8	x	___	=	___
c.	10 x 11	=	___	x	___	=	___
d.	12 x 10	=	___	x	___	=	___
e.	8 x 20	=	___	x	___	=	___
f.	14 x ___	=	10	x	___	=	___
g.	7 x ___	=	13	x	___	=	___
h.	6 x ___	=	20	x	___	=	___

3. Follow the arrows and count by the number in the first box.

a.

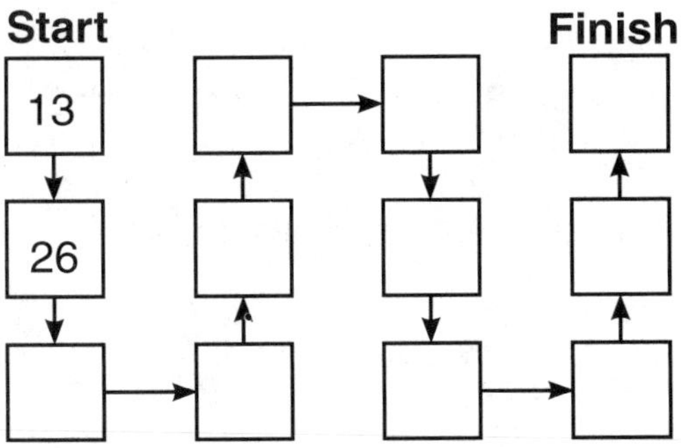

b.

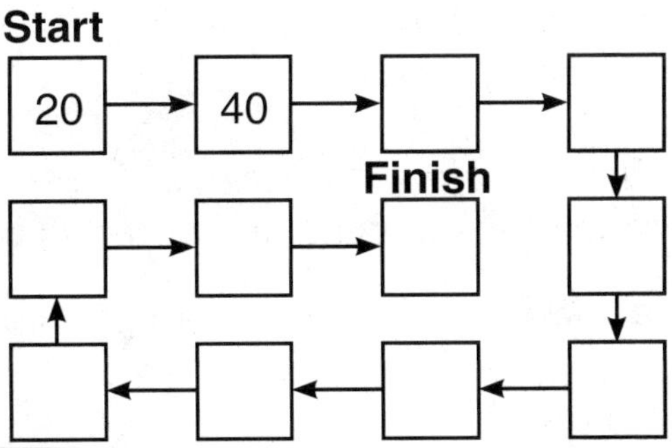

4. Write in the box the product for each pair of factors.

a. 6 and 9 = ☐

b. 13 and 10 = ☐

c. 9 and 11 = ☐

d. 14 and 6 = ☐

e. 7 and 15 = ☐

f. 18 and 3 = ☐

g. 30 and 10 = ☐

h. 16 and 6 = ☐

i. 22 and 2 = ☐

j. 17 and 5 = ☐

k. 21 and 3 = ☐

l. 24 and 5 = ☐

Name **Date**

1. Write the factors for each of the multiples.

a. 15 = ___ , ___ , ___ , and ___

b. 36 = ___ , ___ , ___ , ___ , ___ , ___ , ___ , ___ , and ___

c. 100 = ___ , ___ , ___ , ___ , ___ , ___ , ___ , ___ , and ___

d. 20 = ___ , ___ , ___ , ___ , ___ , and ___

e. 64 = ___ , ___ , ___ , ___ , ___ , ___ , and ___

f. 19 = ___ and ___

2. Color the numbers that are multiples of the number in the circle.

a.

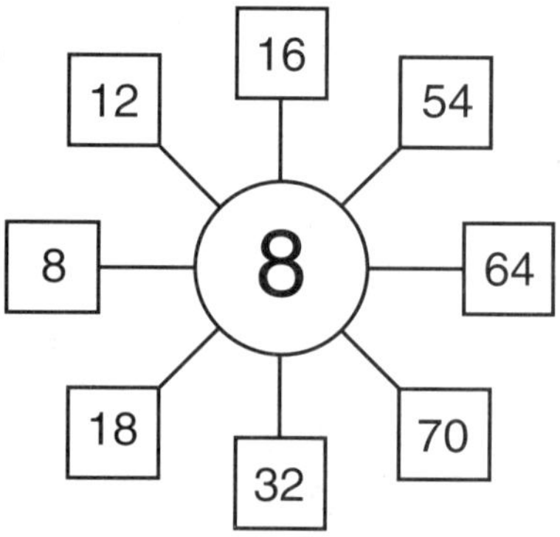

b.

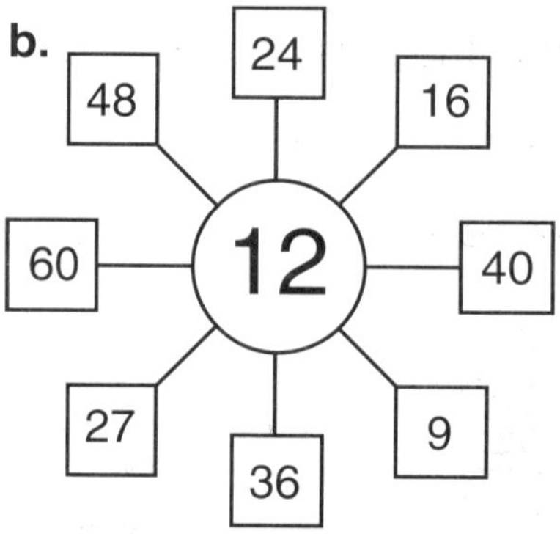

c.

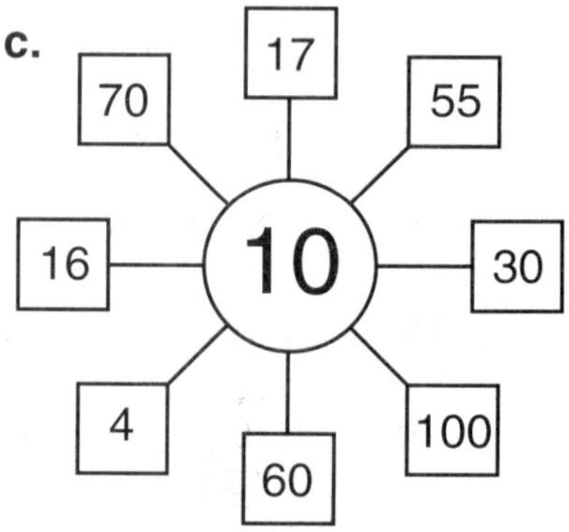

3. Solve the money problems.

a. $1.53 x 6

b. $2.75 x 5

c. $7.93 x 4

d. $8.00 x 9

e. $6.32 x 7

f. $5.04 x 7

g. $3.92 x 6

h. $2.53 x 6

i. $8.16 x 7

j. $6.29 x 6

k. $1.83 x 5

l. $4.62 x 2

4. Watch the parentheses.

a. (11 x 10) + 4 = ___

b. (9 x 9) + 15 = ___

c. (10 x 12) + 16 = ___

d. 19 + (16 x 3) = ___

e. (17 x 3) + 10 = ___

f. (23 x 3) + 11 = ___

g. 20 + (15 x 2) = ___

h. (21 x 2) + 7 = ___

i. (16 x 8) + 0 = ___

j. 10 + (23 x 1) = ___

k. (42 x 4) + 20 = ___

l. (50 x 2) + 20 = ___

Name **Date**

1. Multiplying by 10, 100, and 1,000.

a. 3 x 10 = ________ **b.** 8 x 10 = ________ **c.** 10 x 10 = ________

d. 23 x 10 = ________ **e.** 33 x 10 = ________ **f.** 4 x 100 = ________

g. 16 x 100 = ________ **h.** 49 x 100 = ________ **i.** 56 x 100 = ________

j. 3 x 1,000 = ________ **k.** 14 x 1,000 = ________ **l.** 230 x 1,000 = ________

m. 531 x 10 = ________ **n.** 263 x 100 = ________ **o.** 72 x 1,000 = ________

2. Write the amount of coins needed to make each statement true.

a. $3 is worth the same as _____ 10¢. **b.** $0.70 is worth the same as _____ 10¢.

c. _____ 5¢ coins are worth the same as 95¢. **d.** $1.50 is worth the same as _____ 50¢.

e. $1 is worth the same as _____ 50¢. **f.** $3.00 is worth the same as _____ 50¢.

g. $0.80 is worth the same as _____ 5¢. **h.** $1.75 is worth the same as _____ 5¢.

3. What is the value of each?

a.

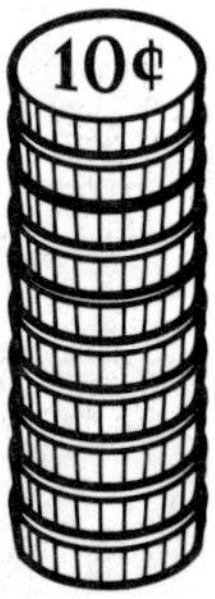

10 coins

____ x ____ = ____

b.

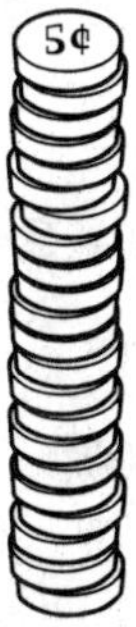

22 coins

____ x ____ = ____

c.

14 coins

____ x ____ = ____

d.

19 coins

____ x ____ = ____

e.

14 coins

____ x ____ = ____

4. Complete the table using your calculator.

	x	2	100	32	8	27	10	12	15	20
a.	7									
b.	8									
c.	9									
d.	10									
e.	16									

Name	Date

1. Multiplying by tens. The first one has been done for you. Work out each problem very carefully.

	H	T	O
a.		3	2
x		2	7
	2	2	4
+	6	4	0
	8	6	4

b. 44 x 16

c. 18 x 22

d. 27 x 23

e. 36 x 28

f. 55 x 46

g. 33 x 41

h. 66 x 17

i. 59 x 24

j. 17 x 22

k. 66 x 18

l. 22 x 33

m. 18 x 17

n. 25 x 15

o. 39 x 58

2. Write *true* or *false*.

a. (4 x 8) + (5 x 8) = 72 ________

b. 9 + 9 + 9 + 9 = 3 x 9 ________

c. 12 x 10 x 2 = 300 ________

d. (6 x 9) + 100 = 172 ________

e. (2 x 10) – (1 x 10) = 10 ________

f. (3 x 36) – 20 = 188 ________

g. (23 x 10) + (23 x 10) = 459 ________

h. 90 x 1,000 = 9,000 ________

3. Answer these questions.

a. What number between 70 and 80 is a multiple of 11? ________

b. What number between 80 and 90 is a multiple of 9? ________

c. What is the largest number that is a factor of 18 and 45? ________

d. Multiply 463 by 1,000. ________

e. Write a multiplication equation using the numbers 17, 5, and 85. ________

f. What is the product of 23 and 48? ________

g. What is the product of 17 and 13? ________

Name	**Date**

1. Find the missing number.

a. 6 x _____ = 120	**b.** 7 x _____ = 49	**c.** 30 x _____ = 300
d. 73 x _____ = 438	**e.** 16 x _____ = 144	**f.** 20 x _____ = 600
g. 25 x _____ = 1,000	**h.** 75 x _____ = 7,500	**i.** 27 x _____ = 540

2. Complete the number wheels.

a.

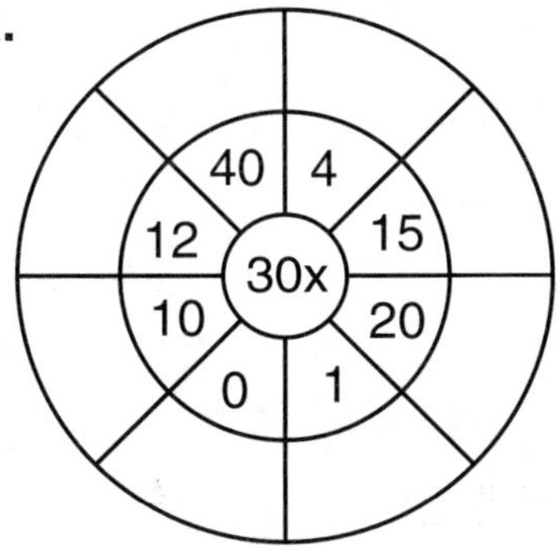

b.

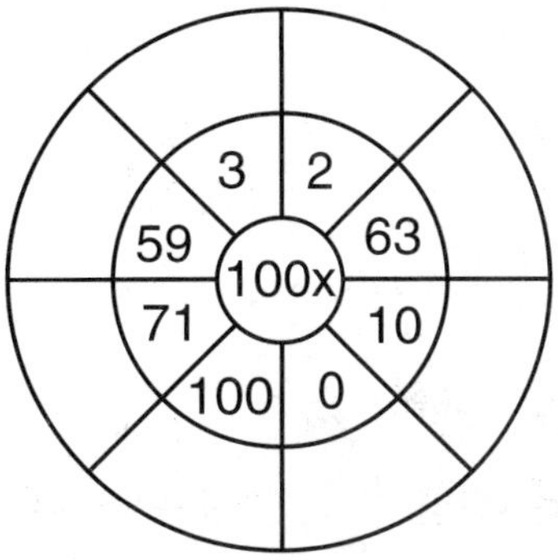

c.

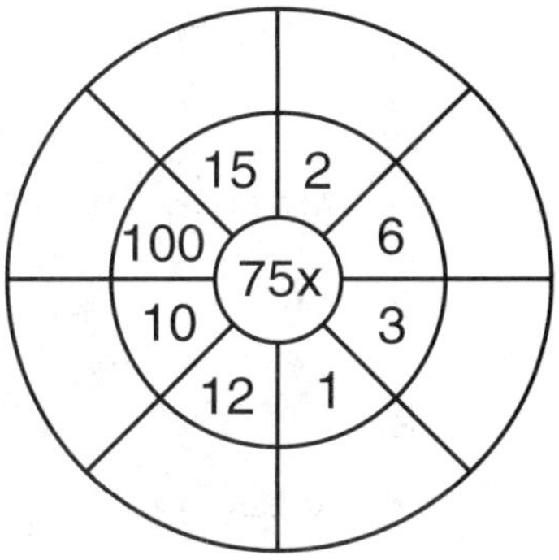

3. Show all your work.

a. 27 x 54	**b.** 18 x 15	**c.** 21 x 19	**d.** 32 x 55	**e.** 67 x 81
f. 87 x 32	**g.** 77 x 61	**h.** 82 x 49	**i.** 76 x 33	**j.** 62 x 59

4. Use your calculator to help you travel from home to school.

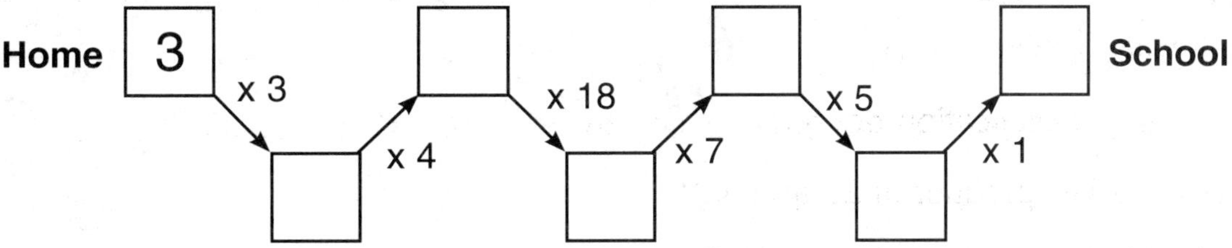

Name **Date**

1. Watch the columns and show all work.

a. 9 x 7

b. 18 x 6

c. 236 x 4

d. 386 x 5

e. 276 x 8

f. 24 x 23

g. 43 x 18

h. 55 x 21

i. 73 x 45

j. 52 x 27

2. **a.** Write the factors of 25. ______________________

b. Write the factors of 100. ______________________

c. What is the product of 23 and 10? ______________________

d. What number between 90 and 100 is a multiple of 12? ______________________

e. (4 x 40) + (5 x 40) = __________

f. (51 x 10) + (51 x 10) = __________

g. (6 x 50) + (0 x 50) = __________

h. (73 x 1,000) + 10 = __________

3. Solve these problems and show your work.

a. Harry bought a dozen plums at 60¢ each and half a dozen peaches for $1.00 each. How much did all of his fruit cost?

(___ x ___) + (___ x ___) = ___

b. Pip bought 3 apples at 15¢ each, 2 bananas at 20¢ each, and 3 watermelons at 70¢ each. How much was her fruit bill?

(___ x ___) + (___ x ___) +

(___ x ___) = ___

4. Complete the number wheels.

a.

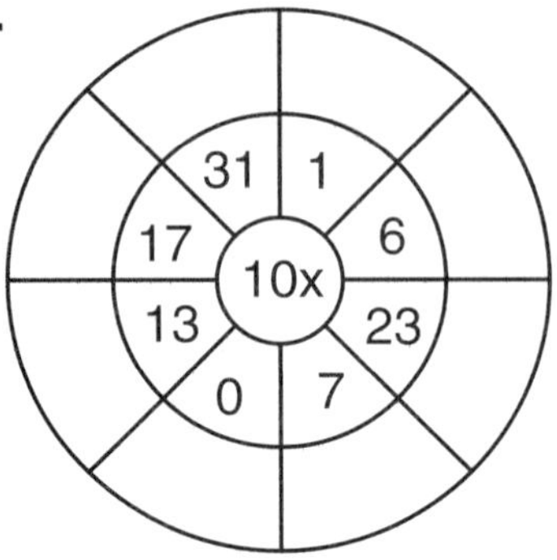

b.

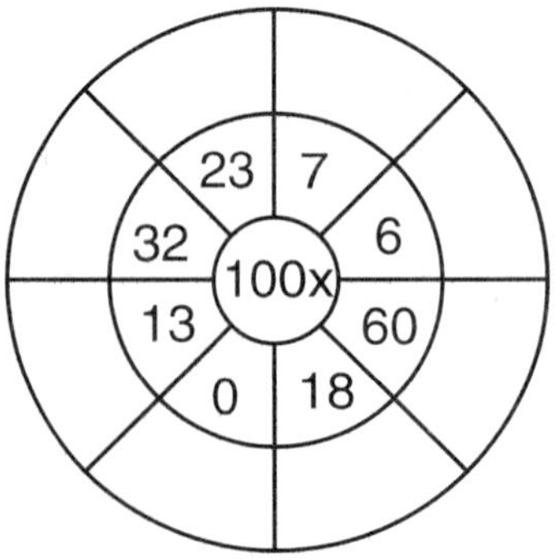

c.

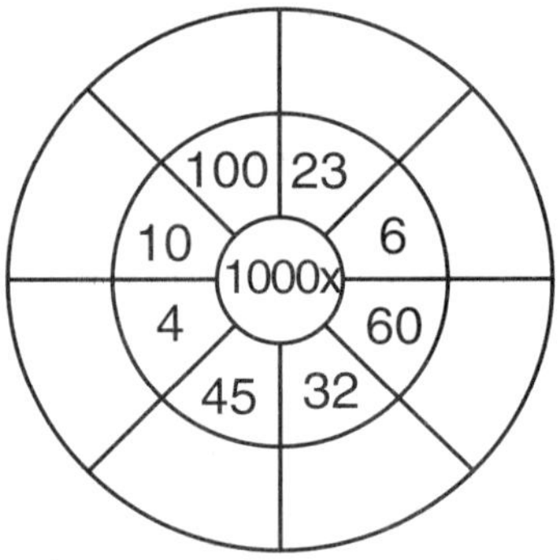

DIVISION

The skills of grouping, sharing, dividing, and calculating are used in these division units. Formal algorithms are worked with and without remainders and extend to three digits divided by a one-digit divisor. Pictorial representation, formal working, number wheels, problem solving, completing grids, patterning, and true/false statements are used to reinforce the division concept. There are also two assessment pages.

DIVISION PRACTICE

Unit 1

Grouping
Remainders
Using correct signs
Problem solving

Objectives

- *manipulate groups of objects by combining and/ or separating*
- *check whether answers to problems are correct and sensible*
- *use a variety of informal strategies to approximate, calculate, and represent solutions to division problems*
- *generate and describe number patterns using a variety of strategies*
- *write complete equivalent number statements using the division signs*
- *use knowledge of the multiplication tables to help solve a division problem*

Language

sharing, groups of, division, sign, remainder, equal, left over, repeated, subtraction, regrouped

Materials/Resources

counters, blank paper, unit cubes (Base 10), colored pencils

Contents of Student Pages

** Materials needed for each reproducible student page*

Remember

Before starting ensure each student:

- ❑ *reads problem solving questions twice.*
- ❑ *checks answers.*
- ❑ *uses a ruler when working out an equation formally.*

Additional Activities

- ❑ *Draw number lines. Start at the large number being divided and then subtract back the number that is shared.*

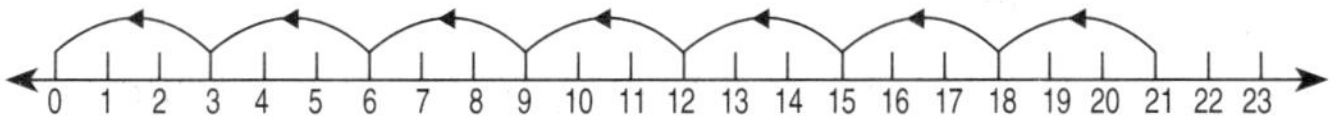

$21 \div 3 = 7$

- ❑ *Use a pack of playing cards to play sharing games.*
- ❑ *Make two dice numbered 4–9. Each player throws the dice and finds the product of the two numbers rolled. The highest score wins. Play in small groups.*

Answers

Page 60 Grouping

1. a. 5, 5 c. 10, 10
 b. 8, 8 d. 9, 9
2. a. 7, 1 b. 26, 6, 2 c. 29, 5, 4
3. a. 4, 2 b. 7, 1

Page 61 Grouping and Remainders

1. a. 6, 6 d. 7, 7 g. 9, 9
 b. 8, 8 e. 9, 9 h. 12, 12
 c. 3, 3 f. 7, 7
2. a. 18 c. 32 e. 27
 b. 30 d. 14 f. 15
3. a. 6 remainder 2
 b. 7 remainder 3
 c. 2 remainder 9
 d. 4 remainder 2

Page 62 Remainders

1. a. 10 ÷ 2 = 5 c. 35 ÷ 7 = 5 e. 12 ÷ 12 = 1
 b. 20 ÷ 4 = 5 d. 16 ÷ 4 = 4 f. 18 ÷ 9 = 2
2. a. 5 f. 3 k. 9
 b. 6 g. 2 l. 3
 c. 6 h. 1 m. 9
 d. 5 i. 7 n. 3
 e. 6 j. 2 o. 3
3. a. 5 r3 f. 5 r2 k. 3 r1
 b. 3 r3 g. 2 r2 l. 1 r1
 c. 2 r1 h. 3 r2 m. 9 r1
 d. 8 r1 i. 2 r2 n. 2 r7
 e. 3 r1 j. 2 r1 o. 4 r3
4. a. 5 r1 e. 5 r6 i. 2
 b. 4 r5 f. 2 r1 j. 5
 c. 2 r2 g. 9 k. 5 r6
 d. 5 r1 h. 7 r1 l. 1 r4

Page 63 Using Signs

1. All answers are ÷ and =.
2. a. 42 ÷ 6 = 7 d. 36 ÷ 6 = 6 g. 45 ÷ 5 = 9
 b. 9 ÷ 3 = 3 e. 10 ÷ 2 = 5 h. 6 ÷ 6 = 1
 c. 21 ÷ 7 = 3 f. 28 ÷ 4 = 7 i. 32 ÷ 8 = 4
3. a. 4 r4 e. 4 r2 i. 4 r3
 b. 1 r2 f. 2 r1 j. 8 r2
 c. 3 r1 g. 2 r1 k. 2 r8
 d. 2 r1 h. 1 r7 l. 7 r1
4. Teacher to check.
5. b. 24 ÷ 6 = 4, 4 x 6 = 24
 c. 33 ÷ 3 = 11, 11 x 3 = 33
 d. 27 ÷ 9 = 3, 3 x 9 = 27
 e. 16 ÷ 8 = 2, 2 x 8 = 16
 f. 22 ÷ 2 = 11, 11 x 2 = 22
 g. 66 ÷ 6 = 11, 11 x 6 = 66
 h. 81 ÷ 9 = 9, 9 x 9 = 81
 i. 35 ÷ 7 = 5, 5 x 7 = 35
 j. 24 ÷ 6 = 4, 4 x 6 = 24

Page 64 Problem Solving

1. a. 24 ÷ 3 = 8
 b. 72 ÷ 9 = 8
 c. 42 ÷ 7 = 6
 d. 82 ÷ 8 = 10 r2
2. Teacher to check problems.
 a. 9
 b. 4
 c. 7
 d. 9
3. a.

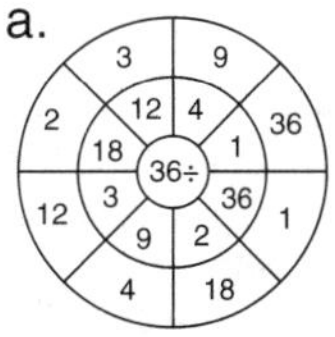

b.

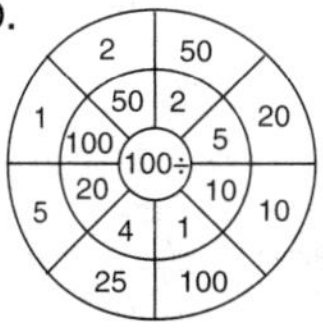

c.

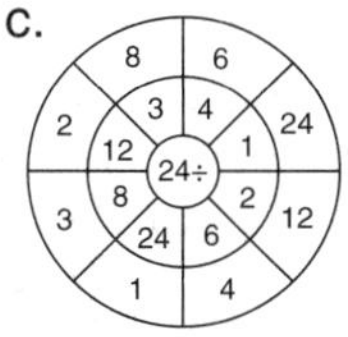

Page 65 Division Equations

1. a. 4 d. 5 g. 7 j. 3
 b. 6 e. 10 h. 6 k. 8
 c. 8 f. 12 i. 4 l. 7
2. a. 4 r1 d. 6 r2 g. 7 r1 j. 7 r1
 b. 6 r2 e. 8 r2 h. 1 r4 k. 8 r1
 c. 2 r1 f. 8 r1 i. 4 r2 l. 4 r5
3. a. 5 d. 4 r1 g. 4 j. 8, 1
 b. 28 e. 3 r4 h. 3 r6 k. 3, 5 or 5, 3
 c. 6 f. 32 i. 3 l. 37
4. Teacher to check.

Page 66 Assessment

1. a. 53 ÷ 7 = 7 r4 b. 31 ÷ 7 = 4 r3
2. a. 4 d. 3 g. 1 r5 j. 6
 b. 4 e. 5 h. 2 k. 4
 c. 4 f. 6 r1 i. 4 r4 l. 8 r1
3. a. 28 ÷ 4 = 7
 b. 300 ÷ 10 = 30
4. a. 24 ÷ 6 = 4, 4 x 6 = 24
 b. 36 ÷ 4 = 9, 9 x 4 = 36
 c. 42 ÷ 7 = 6, 6 x 7 = 42
 d. 21 ÷ 3 = 7, 7 x 3 = 21
 e. 56 ÷ 8 = 7, 7 x 8 = 56
 f. 18 ÷ 9 = 2, 2 x 9 = 18
5. a. 18 r2 e. 4
 b. 26 f. 9 r1
 c. 4 r3 g. 9 r1
 d. 5 r6 h. 10 r2

Name **Date**

1. You will need colored pencils and counters. Use different colors to make equal groups. Complete the word and number sentence to record results.

a.

There are ______ groups of 5 hearts.
25 ÷ 5 = ______

b.

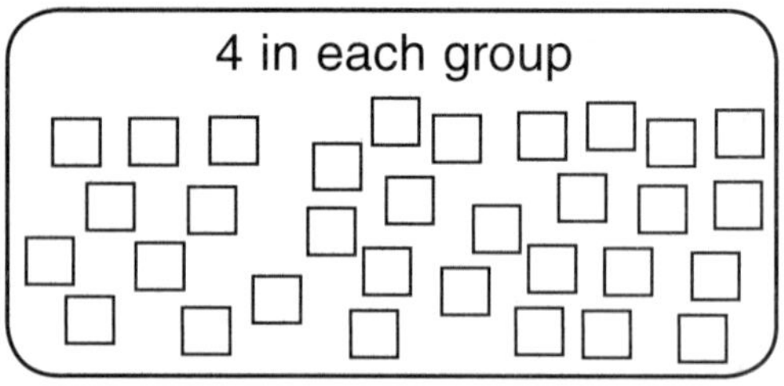

There are ______ groups of 4 squares.
32 ÷ 4 = ______

c.

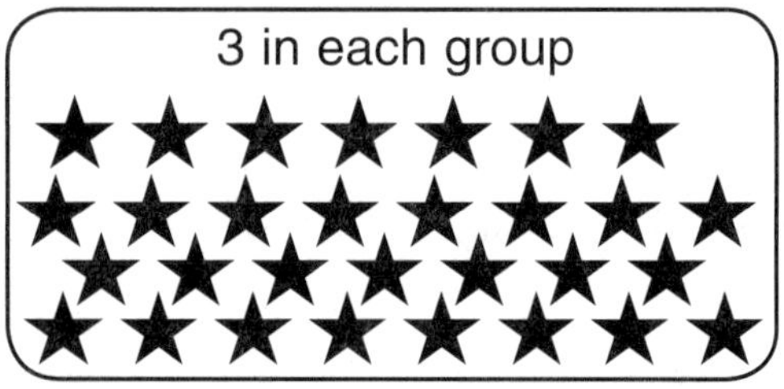

There are ______ groups of 3 stars.
30 ÷ 3 = ______

d.

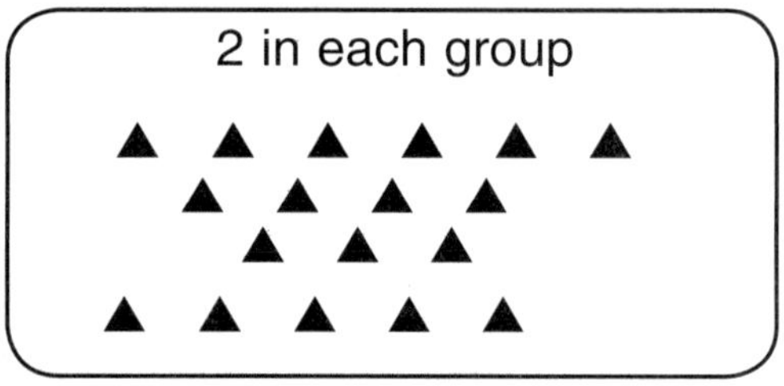

There are ______ groups of 2 triangles.
18 ÷ 2 = ______

2. Find out how much will be left over after the equal shares have been circled. Complete the sentences below.

a.

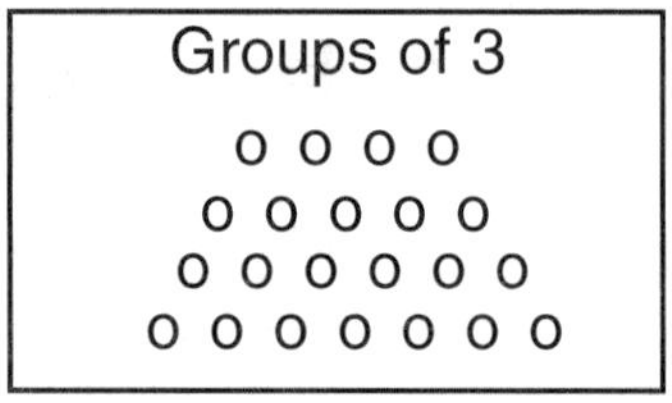

22 shared between 3 equals ______ with ______ left over.

b.

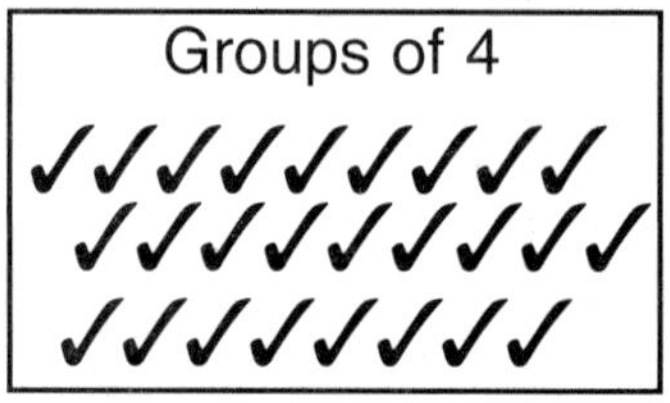

______ shared between 4 equals ______ with ______ left over.

c.

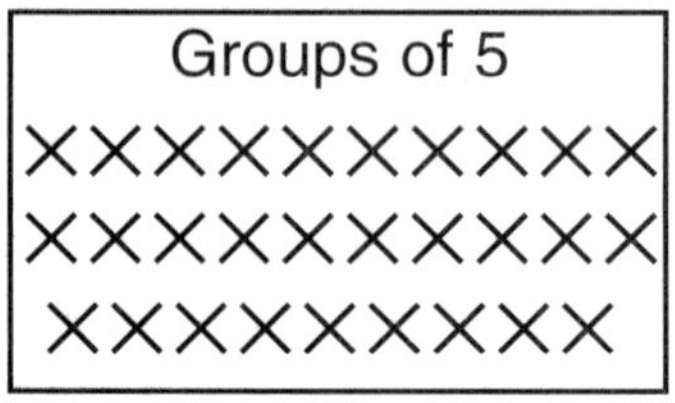

______ shared between 5 equals ______ with ______ left over.

3. Use counters to help you solve these sharing problems.

a. 18 counters shared among 4 people equals ______ remainder ______ .

b. 36 counters shared among 5 people equals ______ remainder ______ .

Name **Date**

You will need counters.

1. Solve these problems. The first one has been done for you.

a. How many groups of 4 in 24? 6 — 24 ÷ 4 = 6

b. How many groups of 5 in 40? ______ 40 ÷ 5 = ______

c. Share 9 pencils among 3 boys. ______ 9 ÷ 3 = ______

d. Share 70 balls among 10 children. ______ 70 ÷ 10 = ______

e. How many groups of 9 in 81? ______ 81 ÷ 9 = ______

f. Share 42 candies with 6 girls. ______ 42 ÷ 6 = ______

g. How many groups of 4 in 36? ______ 36 ÷ 4 = ______

h. How many groups of 2 in 24? ______ 24 ÷ 2 = ______

2. Join groups and remainders together. Record the total before it was divided.

a. 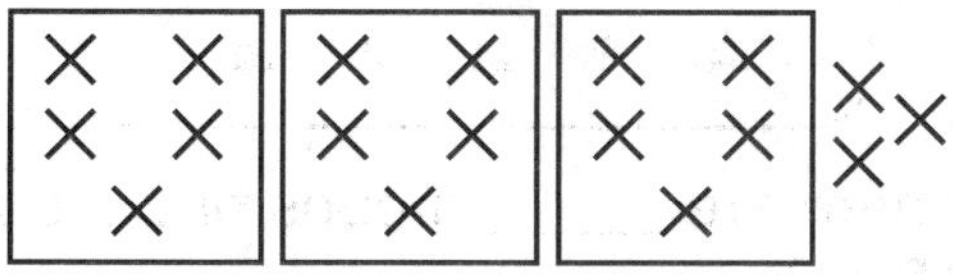 ______

b. 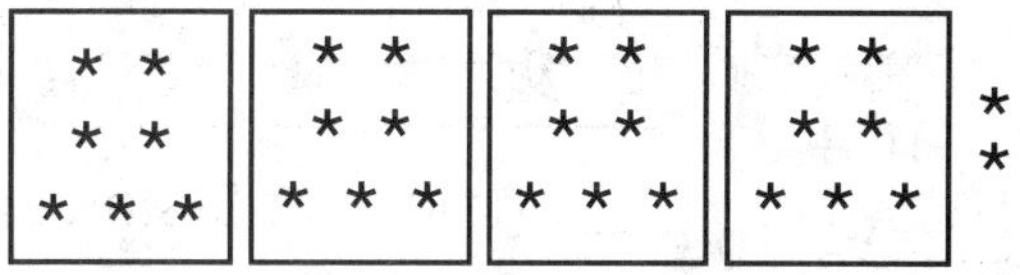 ______

c. 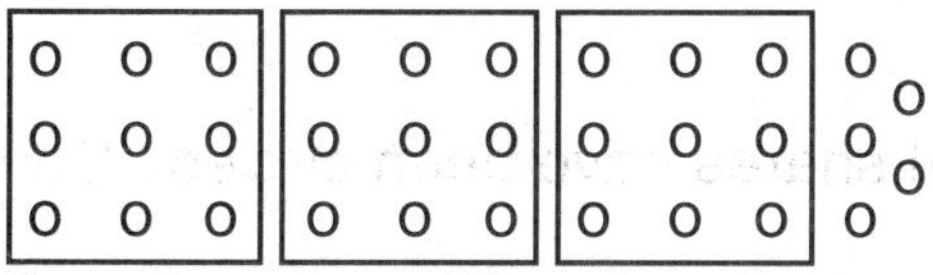 ______

d. 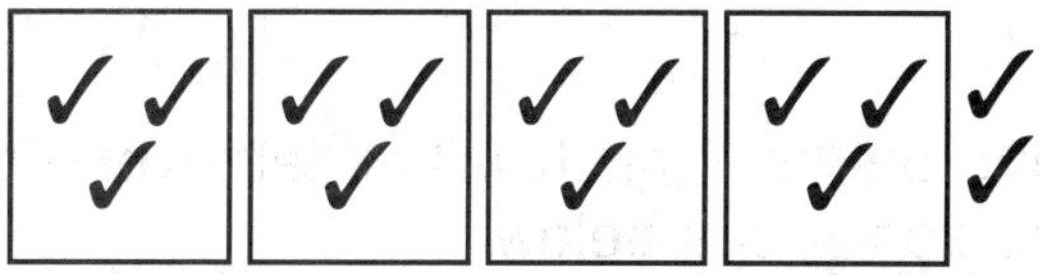 ______

e. 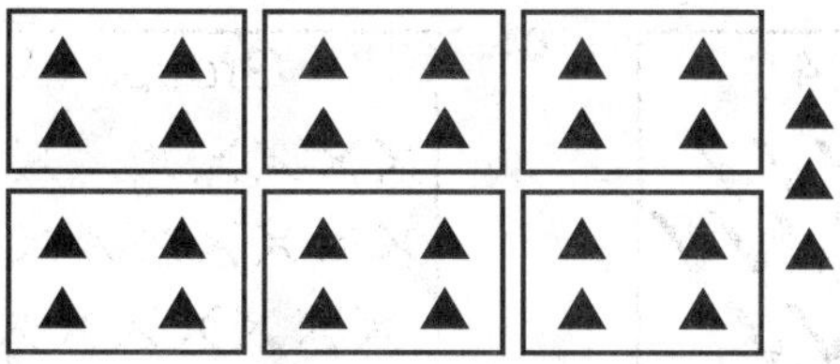 ______

f. 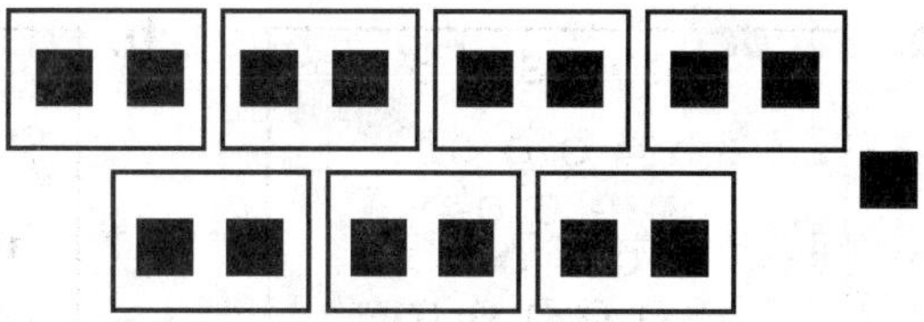 ______

3. Use unit cubes to find out how many groups of 12 there are in each problem. Write the amount and then the remainders.

a. 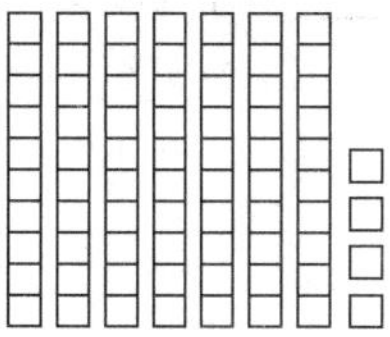 ____ remainder ____

b. 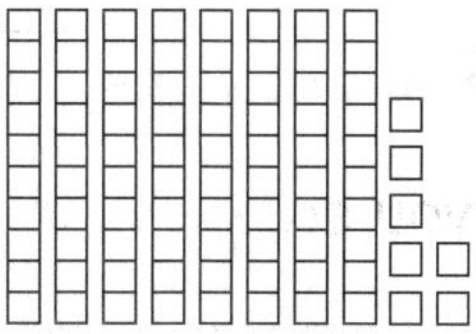 ____ remainder ____

c. 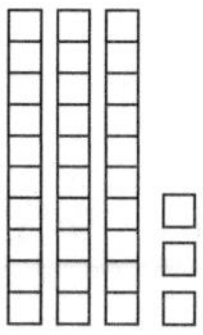 ____ remainder ____

d. 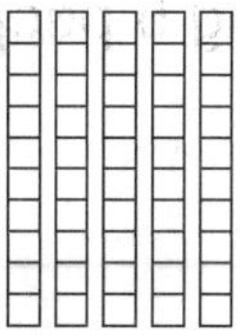 ____ remainder ____

Name **Date**

You will need counters.

1. Read each problem and write the division number sentence using the sign ÷. Use counters to help you solve the problem.

   a. Ten pencils divided into two equal groups. ☐ ÷ ☐ = ☐

   b. Twenty oranges shared into four bowls equally. ☐ ÷ ☐ = ☐

   c. Thirty-five biscuits shared equally between seven children. ☐ ÷ ☐ = ☐

   d. Sixteen books shared equally between four groups. ☐ ÷ ☐ = ☐

   e. Twelve drinks divided between twelve soccer players. ☐ ÷ ☐ = ☐

   f. Eighteen apples shared equally between nine girls. ☐ ÷ ☐ = ☐

2. Use counters to help complete the number sentences.

a. 25 ÷ 5 = ______	b. 30 ÷ 5 = ______	c. 24 ÷ 4 = ______
d. 15 ÷ 3 = ______	e. 12 ÷ 2 = ______	f. 18 ÷ 6 = ______
g. 20 ÷ 10 = ______	h. 10 ÷ 10 = ______	i. 14 ÷ 2 = ______
j. 8 ÷ 4 = ______	k. 27 ÷ 3 = ______	l. 30 ÷ 10 = ______
m. 18 ÷ 2 = ______	n. 21 ÷ 7 = ______	o. 9 ÷ 3 = ______

3. Complete these number sentences and show how many are left over by using "r" for the remainder. The first one has been done for you. Use counters to help you.

a. 28 ÷ 5 = 5 r 3	b. 30 ÷ 9 = ____ r ____	c. 21 ÷ 10 = ____ r ____
d. 17 ÷ 2 = ____ r ____	e. 16 ÷ 5 = ____ r ____	f. 32 ÷ 6 = ____ r ____
g. 12 ÷ 5 = ____ r ____	h. 11 ÷ 3 = ____ r ____	i. 10 ÷ 4 = ____ r ____
j. 15 ÷ 7 = ____ r ____	k. 13 ÷ 4 = ____ r ____	l. 6 ÷ 5 = ____ r ____
m. 19 ÷ 2 = ____ r ____	n. 23 ÷ 8 = ____ r ____	o. 31 ÷ 7 = ____ r ____

4. Complete these division problems. Watch out for those that have remainders. Use counters to help you.

a. 36 ÷ 7 = ______	b. 29 ÷ 6 = ______	c. 12 ÷ 5 = ______
d. 41 ÷ 8 = ______	e. 41 ÷ 7 = ______	f. 11 ÷ 5 = ______
g. 45 ÷ 5 = ______	h. 43 ÷ 6 = ______	i. 10 ÷ 5 = ______
j. 50 ÷ 10 = ______	k. 51 ÷ 9 = ______	l. 9 ÷ 5 = ______

Name **Date**

1. Place the ÷ sign and the = sign in the squares so that each number sentence is correct.

a. 48 [ ] 6 [ ] 8 **b.** 100 [ ] 10 [ ] 10 **c.** 18 [ ] 2 [ ] 9

d. 55 [ ] 5 [ ] 11 **e.** 63 [ ] 9 [ ] 7 **f.** 14 [ ] 7 [ ] 2

g. 25 [ ] 5 [ ] 5 **h.** 56 [ ] 7 [ ] 8 **i.** 36 [ ] 12 [ ] 3

2. Include the division sign, equal sign, and the answer.

a. 42 [ ] 6 [ ] ___ **b.** 9 [ ] 3 [ ] ___ **c.** 21 [ ] 7 [ ] ___

d. 36 [ ] 6 [ ] ___ **e.** 10 [ ] 2 [ ] ___ **f.** 28 [ ] 4 [ ] ___

g. 45 [ ] 5 [ ] ___ **h.** 6 [ ] 6 [ ] ___ **i.** 32 [ ] 8 [ ] ___

3. Solve and watch out for any remainders.

a. 28 ÷ 6 = ____ r ____ **b.** 8 ÷ 6 = ____ r ____ **c.** 28 ÷ 9 = ____ r ____

d. 17 ÷8 = ____ r ____ **e.** 14 ÷ 3 = ____ r ____ **f.** 15 ÷ 7 = ____ r ____

g. 21 ÷ 10 = ____ r ____ **h.** 16 ÷ 9 = ____ r ____ **i.** 23 ÷ 5 = ____ r ____

j. 26 ÷ 3 = ____ r ____ **k.** 28 ÷ 10 = ____ r ____ **l.** 29 ÷ 4 = ____ r ____

4. Make some of your own division number sentences.

a. ___ [ ] ___ [ ] ___ **b.** ___ [ ] ___ [ ] ___ **c.** ___ [ ] ___ [ ] ___

d. ___ [ ] ___ [ ] ___ **e.** ___ [ ] ___ [ ] ___ **f.** ___ [ ] ___ [ ] ___

g. ___ [ ] ___ [ ] ___ **h.** ___ [ ] ___ [ ] ___ **i.** ___ [ ] ___ [ ] ___

5. Use multiplication tables to check that you have divided correctly. The first one has been done for you.

a. 15 ÷ 3 = 5, 5 x 3 = 15 **b.** 24 ÷ 6 = ___ , 6 x ___ = 24

c. 33 ÷ 3 = 11, ___ x ___ = ___ **d.** 27 ÷ 9 = ___ , ___ x ___ = ___

e. 16 ÷ 8 = ___ , ___ x ___ = ___ **f.** 22 ÷ 2 = ___ , ___ x ___ = ___

g. 66 ÷ 6 = ___ , ___ x ___ = ___ **h.** 81 ÷ 9 = 9, ___ x ___ = 81

i. 35 ÷ 7 = ___ , ___ x ___ = ___ **j.** 24 ÷ 6 = ___ , ___ x ___ = ___

Name **Date**

1. Solve these problems.

a. Adam, Nicole, and John had twenty-four cards. How many did each received when they shared them equally?

___ ÷ ___ = ___

b. On an Easter egg hunt, seventy-two chocolate eggs were found. How many did each of the nine children receive when they shared them equally?

___ ÷ ___ = ___

c. Forty-two tomato plants were planted in seven equal rows. How many were planted in each row?

___ ÷ ___ = ___

d. Eighty-two oranges were picked and placed equally into eight boxes. How many oranges were in each box? Were there any left over?

___ ÷ ___ = ___ r ___

2. Write a problem for each number sentence and then write the answer.

a. ______________________________

____

____

____

36 ÷ 4 = ___

b. ______________________________

____

____

____

28 ÷ 7 =

c. ______________________________

____

____

____

56 ÷ 8 = ___

d. ______________________________

____

____

____

45 ÷ 5 = ___

3. Complete.

a.

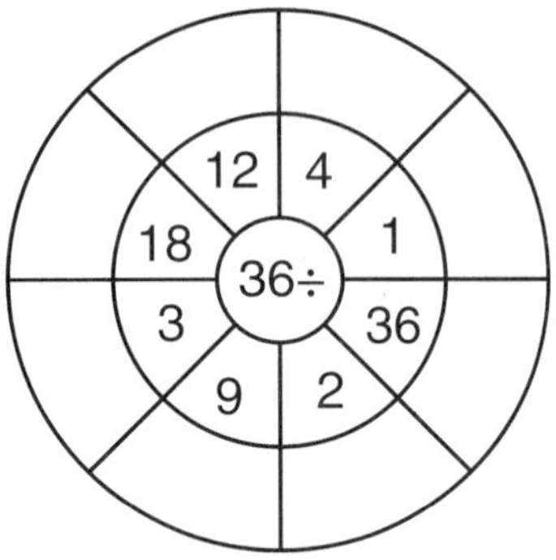

b.

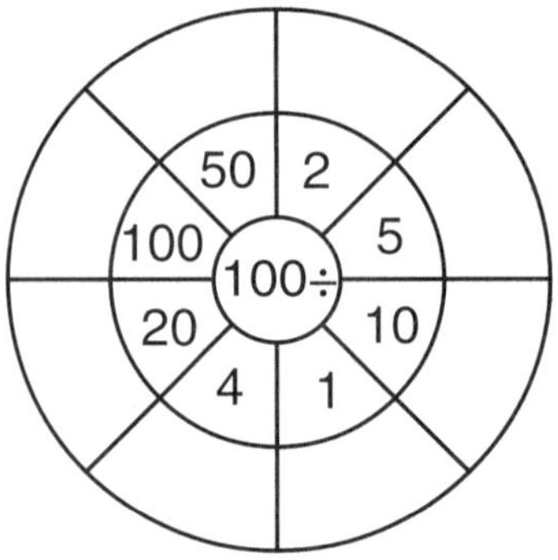

c.

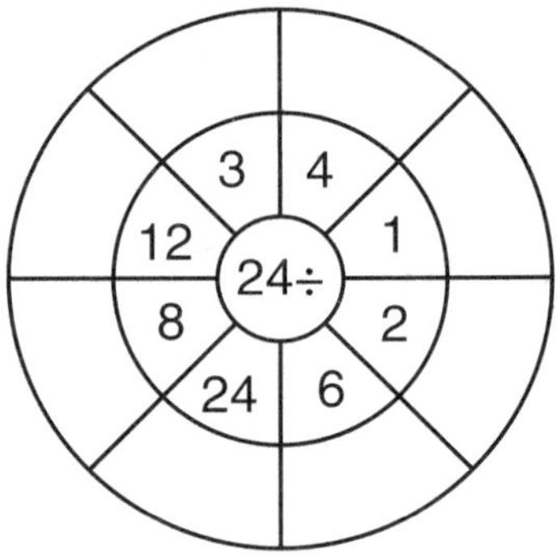

Name **Date**

1. Complete the following. The first one has been done for you.

a. $4\overline{)16}$ = 4	**b.** $2\overline{)12}$	**c.** $3\overline{)24}$	**d.** $4\overline{)20}$
e. $7\overline{)70}$	**f.** $5\overline{)60}$	**g.** $4\overline{)28}$	**h.** $5\overline{)30}$
i. $8\overline{)32}$	**j.** $6\overline{)18}$	**k.** $7\overline{)56}$	**l.** $6\overline{)42}$

2. Watch out for remainders. The first one has been done for you.

a. $6\overline{)25}$ = 4 r 1	**b.** $5\overline{)32}$	**c.** $7\overline{)15}$	**d.** $4\overline{)26}$
e. $5\overline{)42}$	**f.** $6\overline{)49}$	**g.** $7\overline{)50}$	**h.** $8\overline{)12}$
i. $3\overline{)14}$	**j.** $2\overline{)15}$	**k.** $3\overline{)25}$	**l.** $6\overline{)29}$

3. Fill in the missing numbers so that the equation is true.

a. $\square\overline{)25}$ = 5	**b.** $7\overline{)\square\square}$ = 4	**c.** $\square\overline{)36}$ = 6	**d.** $5\overline{)21}$ = $\square$ r $\square$
e. $7\overline{)25}$ = $\square$ r $\square$	**f.** $8\overline{)\square\square}$ = 4	**g.** $\square\overline{)40}$ = 10	**h.** $7\overline{)27}$ = $\square$ r $\square$
i. $\square\overline{)36}$ = 12	**j.** $\square\overline{)41}$ = 5 r $\square$	**k.** $\square\overline{)19}$ = $\square$ r 4	**l.** $3\overline{)\square\square}$ = 1 2 r 1

4. Write some of your own number questions. Give them to a friend to solve.

a. $\overline{)\quad}$	**b.** $\overline{)\quad}$	**c.** $\overline{)\quad}$	**d.** $\overline{)\quad}$
e. $\overline{)\quad}$	**f.** $\overline{)\quad}$	**g.** $\overline{)\quad}$	**h.** $\overline{)\quad}$
i. $\overline{)\quad}$	**j.** $\overline{)\quad}$	**k.** $\overline{)\quad}$	**l.** $\overline{)\quad}$

Name **Date**

1. How many groups of seven are there in each box? Circle each group and write the number sentence below.

a.

53 ÷ ___ = ___ r ___

b.

31 ÷ ___ = ___ r ___

2. Complete.

a. 12 ÷ 3 = ______ **b.** 20 ÷ 5 = ______ **c.** 16 ÷ 4 = ______ **d.** 30 ÷ 10 = ______

e. 25 ÷ 5 = ______ **f.** 19 ÷ 3 = ______ **g.** 11 ÷ 6 = ______ **h.** 10 ÷ 5 = ______

i. 40 ÷ 9 = ______ **j.** 36 ÷ 6 = ______ **k.** 32 ÷ 8 = ______ **l.** 17 ÷ 2 = ______

3. Read the division problems and write the number sentences for each one to help you solve them.

a. There were twenty-eight presents under the Christmas tree for the four Neale children. How many presents did each of the children receive when they shared them equally?

___ ÷ ___ = ___

b. Ten people shared a $300 lotto win equally. How much money did each person receive?

___ ÷ ___ = ___

4. Show another way of writing the division problem using multiplication.

a. 24 ÷ 6 = ___ , ___ x ___ = ___ **b.** 36 ÷ 4 = ___ , ___ x ___ = ___

c. 42 ÷ 7 = ___ , ___ x ___ = ___ **d.** 21 ÷ 3 = ___ , ___ x ___ = ___

e. 56 ÷ 8 = ___ , ___ x ___ = ___ **f.** 18 ÷ 9 = ___ , ___ x ___ = ___

5. Complete these division problems. Show all work.

a. 3)56 **b.** 2)52 **c.** 4)19 **d.** 10)56

e. 9)36 **f.** 3)28 **g.** 5)46 **h.** 6)62

MORE DIVISION PRACTICE

Unit 2

Problem solving
Formal work
Parentheses
Checking by multiplication

Objectives

- *understand the concept of remainders*
- *recall division facts up to 100 and use them in everyday situations*
- *formally record solutions to division problems using division symbols*
- *check whether answers to problems are correct and sensible*

Language

sharing, groups of, division, sign, remainder, grouping, equal, formal, working out

Materials/Resources

counters, blank paper

Contents of Student Pages

** Materials needed for each reproducible student page*

Remember

Before starting ensure each student:

- ❑ *reads problem solving questions twice.*
- ❑ *checks answers.*
- ❑ *uses a ruler when working problems out formally.*

Additional Activity

- *Divide a class of students into two equal groups. One group has calculators and the other doesn't. Give division questions and have a race to see which group answers the question the fastest and has the most correct answers.*

Answers

Page 69 Problem Solving

1. a. 24
   3)72
   72 ÷ 3 = 24

   b. 54
   2)108
   108 ÷ 2 = 54

   c. 26
   7)182
   182 ÷ 7 = 26

   d. $13.83 r 2¢
   6)$83.00
   $83 ÷ 6 = $13.83 remainder 2¢

2. a. 6 r5 | e. 3 r17 | i. 4 r1
   b. 9 r2 | f. 6 | j. 7 r5
   c. 5 r1 | g. 4 r9 | k. 6 r5
   d. 6 r11 | h. 4 r16 | l. 23 r1

Page 70 Show Your Work!

1. a. 67 | g. 11 r4 | l. 13
   b. 14 | h. 44 r1 | m. 13 r2
   c. 142 r2 | i. 207 | n. 120 r1
   d. 13 | j. 137 r3 | o. 30
   e. 140 | k. 130 | p. 186
   f. 12 r3
2. Teacher to check.
3. a. 18, 1 | c. 7, 1 | e. 15, 3
   b. 17, 3 | d. 8, 3

Page 71 Solve It!

1. a. 16 | e. 36 | h. 108
   b. 8 | f. 16 | i. 106
   c. 72 | g. 1 | j. 179
   d. 42
2. a. 66 ÷ 6 = 11, 6 x 11 = 66
   b. 45 ÷ 9 = 5, 5 x 9 = 45
   c. 99 ÷ 33 = 3, 3 x 33 = 99
   d. 135 ÷ 5 = 27, 27 x 5 = 135
   e. 200 ÷ 10 = 20, 20 x 10 = 200
   f. 88 ÷ 8 = 11, 11 x 8 = 88
   g. 106 ÷ 2 = 53, 53 x 2 = 106
   h. 333 ÷ 3 = 111, 111 x 3 = 333
   i. 125 ÷ 5 = 25, 25 x 5 = 125
   j. 45 ÷ 15 = 3, 3 x 15 = 45
3. a. 6
   30)180
   180 ÷ 30 = 6

   b. 89
   2)178
   178 ÷ 2 = 89

   c. 100
   5)500
   500 ÷ 5 = 100

   d. 31
   5)155
   155 ÷ 5 = 31

Page 72 Division Equations

1. a. 34 r1 | g. 121 r1 | l. 101
   b. 147 | h. 139 | m. 110 r1
   c. 168 r1 | i. 133 r6 | n. 26 r3
   d. 197 r1 | j. 90 | o. 134 r3
   e. 141 r2 | k. 114 | p. 127 r4
   f. 138 r3
2. a. True | e. True | h. False
   b. False | f. False | i. False
   c. True | g. False | j. True
   d. False
3. a.
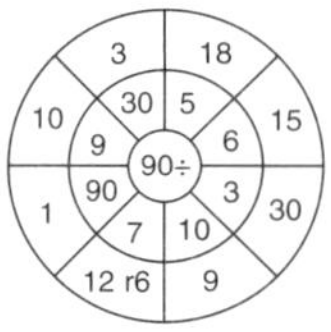

   b.
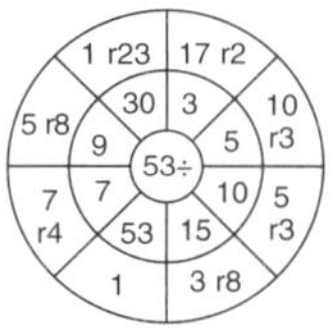

Page 73 Assessment

1. a. 122
   4)488
   488 ÷ 4 = 122 stamps

   b. 250
   8)2,000
   $2,000 ÷ 8 = $250.00 each
2. a. 15 | d. 15 | g. 70
   b. 10 | e. 60 | h. 81
   c. 36 | f. 2
3. a. 130 r1 | e. 94 r4 | i. 100
   b. 132 r2 | f. 51 | j. 152
   c. 131 | g. 70 | k. 141 r3
   d. 127 r4 | h. 104 | l. 126 r1
4. 24 r1
   6)145
   145 ÷ 6 = 24 r1

Name **Date**

1. Read the following division problems and write two number sentences to show your work.

   a. There were 72 sailors and 3 yachts. How many sailors were on each yacht?

   $\overline{)\quad\quad}$ ____ ÷ ____ = ____

   b. There were 108 books to be placed on 2 shelves in the library. How many books were placed on each shelf?

   $\overline{)\quad\quad}$ ____ ÷ ____ = ____

   c. There were 182 children to be organized into 7 equal classes. How many children were in each class?

   $\overline{)\quad\quad}$ ____ ÷ ____ = ____

   d. Eighty-three dollars are to be shared between six children fairly. How much does each child receive and will there be any left over?

   $\overline{)\quad\quad}$ ____ ÷ ____ = ____

2. Solve. Watch for remainders. Use counters and grouping to help.

a. 83 ÷ 13 = ____	**b.** 92 ÷ 10 = ____	**c.** 76 ÷ 15 = ____	**d.** 83 ÷ 12 = ____
e. 71 ÷ 18 = ____	**f.** 66 ÷ 11 = ____	**g.** 93 ÷ 21 = ____	**h.** 84 ÷ 17 = ____
i. 69 ÷ 17 = ____	**j.** 89 ÷ 12 = ____	**k.** 101 ÷ 16 = ____	**l.** 93 ÷ 4 = ____

Name **Date**

1. Solve the problems. The first one has been done for you.

a. $3\overline{)2\,0^{2}\,1}$ = 6 7	b. $4\overline{)56}$	c. $5\overline{)712}$	d. $6\overline{)78}$
e. $7\overline{)980}$	f. $8\overline{)99}$	g. $9\overline{)103}$	h. $2\overline{)89}$
i. $3\overline{)621}$	j. $4\overline{)551}$	k. $5\overline{)650}$	l. $6\overline{)78}$
m. $7\overline{)93}$	n. $8\overline{)961}$	o. $9\overline{)270}$	p. $2\overline{)372}$

2. Create your own division problems and ask a friend to solve them. Compare your answer to check results.

a. $\overline{)\quad}$	b. $\overline{)\quad}$	c. $\overline{)\quad}$	d. $\overline{)\quad}$
e. $\overline{)\quad}$	f. $\overline{)\quad}$	g. $\overline{)\quad}$	h. $\overline{)\quad}$
i. $\overline{)\quad}$	j. $\overline{)\quad}$	k. $\overline{)\quad}$	l. $\overline{)\quad}$

3. Solve the following division problems. Use another piece of paper to show your work.

a. 73 is ______ equal shares of 4 and ______ left over.

b. 88 is ______ equal shares of 5 and ______ left over.

c. 64 is ______ equal shares of 9 and ______ left over.

d. 59 is ______ equal shares of 7 and ______ left over.

e. 93 is ______ equal shares of 6 and ______ left over.

Name **Date**

1. Solve these number problems. Remember to complete the parentheses first.

a. (63 ÷ 7) + (56 ÷ 8) = ____

b. (100 ÷ 10) – (24 ÷ 12) = ____

c. (24 ÷ 3) x (81 ÷ 9) = ____

d. (63 ÷ 3) + (63 ÷ 3) = ____

e. (18 ÷ 2) x (36 ÷ 9) = ____

f. (72 ÷ 9) + (56 ÷ 7) = ____

g. (35 ÷ 5) – (30 ÷ 5) = ____

h. (120 ÷ 10) x (90 ÷ 10) = ____

i. (54 ÷ 9) + 100 = ____

j. 200 – (63 ÷ 3) = ____

2. Use multiplication to check that you have divided correctly.

a. 66 ÷ 6 = ____, 6 x ____ = ____

b. 45 ÷ 9 = ____, 9 x ____ = ____

c. 99 ÷ 33 = ____, 33 x ____ = ____

d. 135 ÷ 5 = ____, 5 x ____ = ____

e. 200 ÷ 10 = ____, 10 x ____ = ____

f. 88 ÷ 8 = ____, ____ x ____ = ____

g. 106 ÷ 2 = ____, ____ x ____ = ____

h. 333 ÷ 3 = ____, ____ x ____ = ____

i. 125 ÷ 5 = ____, ____ x ____ = ____

j. 45 ÷ 15 = ____, ____ x ____ = ____

3. Read the following division problems. Write two ways of working out your answer.

a. Toby had 180 minutes to complete the 30 questions in his math exam. How much time did he need to allow each question so that he would finish?

____ ÷ ____ = ____

b. Jodie had to travel 178 miles to reach her vacation house. If she took 2 days to travel, how far would she travel each day?

____ ÷ ____ = ____

c. Eric won $1,000.00 in a competition. He kept $500.00 and shared the rest of his winnings between his 5 brothers. How much did each brother receive?

____ ÷ ____ = ____

d. Eddy planted 155 plants in 5 rows. How many plants were in each row?

____ ÷ ____ = ____

Name	**Date**

1. Solve the number problems.

a. $3\overline{)1\,0\,3}$ **b.** $3\overline{)4\,4\,1}$ **c.** $4\overline{)6\,7\,3}$ **d.** $4\overline{)7\,8\,9}$

e. $5\overline{)7\,0\,7}$ **f.** $5\overline{)6\,9\,3}$ **g.** $6\overline{)7\,2\,7}$ **h.** $6\overline{)8\,3\,4}$

i. $7\overline{)9\,3\,7}$ **j.** $7\overline{)6\,3\,0}$ **k.** $8\overline{)9\,1\,2}$ **l.** $8\overline{)8\,0\,8}$

m. $9\overline{)9\,9\,1}$ **n.** $9\overline{)2\,3\,7}$ **o.** $5\overline{)6\,7\,3}$ **p.** $7\overline{)8\,9\,3}$

2. Write *true* or *false* for each statement.

a. $63 \div 7 < 60 \div 6$ ______________ **b.** $8 \div 4 > 16 \div 2$ ______________

c. $70 \div 10 = 63 \div 9$ ______________ **d.** $24 \div 4 < 36 \div 6$ ______________

e. $55 \div 5 > 70 \div 7$ ______________ **f.** $44 \div 4 < 10 \div 2$ ______________

g. $36 \div 9 < 24 \div 12$ ______________ **h.** $110 \div 10 = 72 \div 9$ ______________

i. $81 \div 9 < 90 \div 10$ ______________ **j.** $35 \div 7 = 25 \div 5$ ______________

3. Complete the division number wheels.

a.

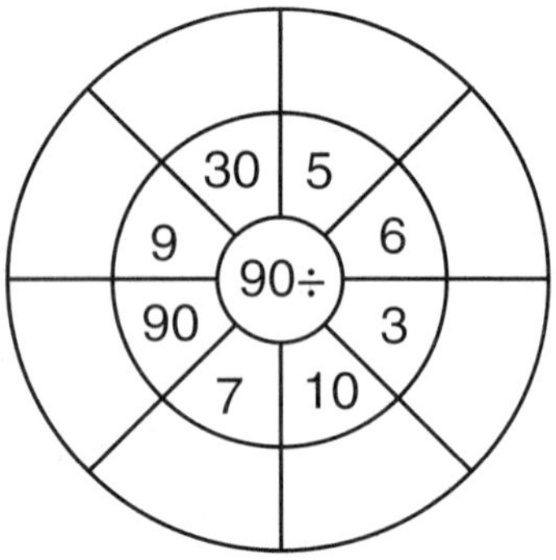

b.

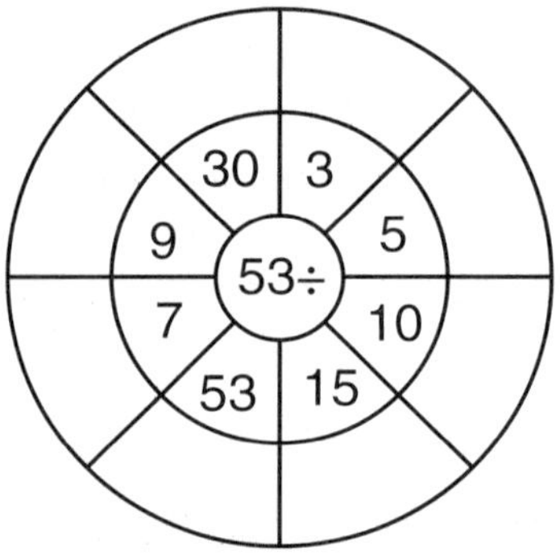

Name **Date**

1. Read the problems and show two ways of solving them.

a. There were 488 stamps altogether when 4 boys put their collections in one album. They had all collected the same amount of stamps. How many did each have before joining them together?

)‾‾‾‾‾‾ ____ ÷ ____ = ____

b. There were 8 girls on the rowing team. How much did each girl win when they shared the prize money of $2,000.00?

)‾‾‾‾‾‾ ____ ÷ ____ = ____

2. Solve the following:

a. (8 ÷ 2) + (33 ÷ 3) = ____

b. (25 ÷ 5) x (20 ÷ 10) = ____

c. (36 ÷ 6) x (36 ÷ 6) = ____

d. (49 ÷ 7) + (56 ÷ 7) = ____

e. (100 ÷ 10) x (24 ÷ 4) = ____

f. (55 ÷ 11) – (33 ÷ 11) = ____

g. (42 ÷ 6) x (60 ÷ 6) = ____

h. (81 ÷ 9) x (81 ÷ 9) = ____

3. Show all work, including the remainders.

a. 6)781 **b.** 4)530 **c.** 5)655 **d.** 7)893

e. 8)756 **f.** 6)306 **g.** 7)490 **h.** 8)832

i. 9)900 **j.** 3)456 **k.** 4)567 **l.** 5)631

4. Circle groups of 6. Record the shares and remainders below in two different ways.

)‾‾‾‾‾‾ ____ ÷ ____ = ____

NUMBER FACTS

Quick recall of number facts is an important asset to all work in mathematics. In these units, this recall is encouraged by working algorithms, solving problems, completing number wheels, patterns and magic squares, finding perimeters, deciding on correct operational signs, and completing a cross number puzzle. Problem solving involves word problems using money, time, and puzzles. There are two assessment pages.

NUMBER FACTS PRACTICE

Unit 1

Problem solving
Correct signs
Operating with 10
Money
< and >

Objectives

- *use number skills involving whole numbers to solve problems*
- *use a variety of strategies to approximate, calculate, and represent solutions to multiplication, division, addition, and subtraction problems*
- *comprehend and solve written word problems*
- *demonstrate an understanding of the monetary value of coins by using them in practical situations*
- *use one or more strategies to solve mathematical problems*

Language

subtraction, addition, division, multiplication, estimating, groups, check, operation, sign, parentheses, perimeter, area, greater than, less than, equal

Materials/Resources

blank paper, colored pencils

Contents of Student Pages

* *Materials needed for each reproducible student page*

Page 77 Mixed Operations
solving addition, subtraction, multiplication, division, number, and word problems

Page 78 Estimating and Checking
solving word and money problems, choosing correct signs

Page 79 Parentheses and Factors
solving number problems with parentheses, solving money problems, and listing factors

Page 80 Perimeter and Area
using greater than and less than signs, completing magic squares, and finding the perimeter and area of shapes

* *colored pencils*

Page 81 Problem Solving
completing number problems and solving word problem

Page 82 Assessment

Remember

Before starting ensure each student:

- ❑ *understands that spacing between each question is important.*
- ❑ *understands the importance of working out the problem neatly.*
- ❑ *develops the habit of checking.*
- ❑ *is reminded to watch for the various operation signs.*

Additional Activities

- *Students make number fact crossnumber puzzles.*
- *Students create number wheels and have partners solve them.*
- *A sum is written on the board. Two players race from the same place to write the answer on the board.*

Answers

Page 77 Mixed Operations

1. a. 89 e. 128 i. 181
   b. 53 f. 26 j. 511
   c. 115 g. 23 k. 325
   d. 64 h. 93 l. 330
2. a. 21, 7, 70, 58, 63
   b. 60, 10, 17, 30, 22
   c. 20, 35, 5, 40, 5
   d. 5, 25, 50, 10, 26
3. a. 7 x 35¢ = $2.45
   b. $4.00 ÷ 8 = $0.50 or 50¢
   c. $10.00 – $6.50 = $3.50
   d. $3.50 + $1.75 + $2.20 = $7.45

Page 78 Estimating and Checking

1. a. 32 ÷ 8 = 4 b. 25 apples c. $3.20
2. a. $3.60 e. $14.01 i. $2.57
   b. $1.50 f. $2.23 j. $4.30
   c. $14.10 g. $13.70 k. $12.07
   d. $3.33 h. $13.98 l. $6.50
3. a. 20 – 5 = 15 d. 8 x 7 = 56 g. 60 – 30 = 30
   b. 30 ÷ 6 = 5 e. 90 ÷ 9 = 10 h. 45 ÷ 5 = 9
   c. 27 – 5 = 22 f. 42 ÷ 6 = 7 i. 23 + 23 = 46

Page 79 Parentheses and Factors

1. a. 7 e. 7 i. 63
   b. 13 f. 40 j. 72
   c. 9 g. 15 k. 10
   d. 52 h. 102 l. 10
2. a. $1.25 + 20¢ = $1.45, $2.00 – $1.45 = $0.55 or 55¢
   b. 15¢ + 10¢ + $1.25 + 20¢ = $1.70, $2.00 – $1.70 = $0.30 or 30¢
   c. 15¢ + 15¢ = 30¢, 40¢ – 30¢ = 10¢
   d. 10¢ + $1.25 = $1.35, $2.20 – $1.35 = $0.85 or 85¢
   e. 20¢ + 20¢ = 40¢, 80¢ – 40¢ = 40¢
   f. 10¢, 20¢ – 10¢ = 10¢
   g. 15¢ + 15¢ + 15¢ = 45¢, 60¢ – 45¢ = 15¢
3. a. 1, 45, 3, 15, 5, 9
   b. 1, 28, 2, 14, 4, 7
   c. 1, 8, 2, 4
   d. 1, 50, 2, 25, 5, 10
4. 56

Page 80 Perimeter and Area

1. a. < d. < g. <
   b. < e. > h. >
   c. < f. >
2. a.

14	16	30
1	15	16
15	31	46

   b.

65	14	51
23	7	16
42	7	35

   c.

2	3	6
4	3	12
8	9	72

3. a. 21 in. c. 44 cm
   b. 21 m d. shade red: shape c
4. a. 36 yd.^{2} c. 75 m^2
   b. 48 cm^2 d. shade yellow: shape a

Page 81 Problem Solving

1. a. 581 e. 71 i. 1,462
   b. 404 f. 1,478 j. 261
   c. 4,404 g. 1,198 k. 1,696
   d. 80 h. 378 l. 101
2. a. (2 x 45¢) + (2 x $1.20) + (2 x 30¢) = $3.90
   b. 25¢
   c. 15 + 32 + 26 = 73 shells
   d. 24 – 5 = 19 apples
3. a. 674
   b. 535
   c. 290
   d. 579
   e. 391
   f. 251

Page 82 Assessment

1. a. $5.65
   b. 534
   c. 61
   d. 2,592
   e. 156
   f. 3,185
   g. 234
   h. 1,637
2. a. 33
   b. 773
   c. 4
   d. 876
   e. 8 r4
   f. 62
3. a. 161 ÷ 7 = 23 miles
   b. (6 x 30¢) + (4 x 20¢) = $2.60 spent, $5.00 – $2.60 = $2.40 change
   c. 13 + 12 + 15 + 10 + 5 = 55 hrs.
4. a. 20 in., 16 in.^{2}
   b. 20 m, 25 m^2
   c. 26 cm, 30 cm^2
5. 24, 6, 36, 56, 41, 50

Name **Date**

1. Complete the number problems. Look carefully at the operation signs.

a. $\begin{array}{r} 73 \\ +\ 16 \\ \hline \end{array}$	**b.** $\begin{array}{r} 89 \\ -\ 36 \\ \hline \end{array}$	**c.** $\begin{array}{r} 23 \\ \times\ 5 \\ \hline \end{array}$	**d.** $\begin{array}{r} 29 \\ +\ 35 \\ \hline \end{array}$
e. $\begin{array}{r} 16 \\ \times\ 8 \\ \hline \end{array}$	**f.** $\begin{array}{r} 53 \\ -\ 27 \\ \hline \end{array}$	**g.** $\begin{array}{r} 62 \\ -\ 39 \\ \hline \end{array}$	**h.** $\begin{array}{r} 14 \\ +\ 79 \\ \hline \end{array}$
i. $\begin{array}{r} 88 \\ +\ 93 \\ \hline \end{array}$	**j.** $\begin{array}{r} 73 \\ \times\ 7 \\ \hline \end{array}$	**k.** $\begin{array}{r} 65 \\ \times\ 5 \\ \hline \end{array}$	**l.** $\begin{array}{r} 304 \\ +\ 26 \\ \hline \end{array}$

2. Follow the arrows and signs to find the answer.

a. 16 —+ 5→ ☐ —÷ 3→ ☐ —x 10→ ☐ —− 12→ ☐ —+ 5→ ☐

b. 52 —+ 8→ ☐ —÷ 6→ ☐ —+ 7→ ☐ —+ 13→ ☐ —− 8→ ☐

c. 33 —− 13→ ☐ —+ 15→ ☐ —÷ 7→ ☐ —x 8→ ☐ —÷ 8→ ☐

d. 20 —÷ 4→ ☐ —x 5→ ☐ —+ 25→ ☐ —− 40→ ☐ —+ 16→ ☐

3. Read the problems and write a numbers sentence to solve each one.

a. Jan bought 7 drinks at 35¢ each. Find the total cost.

____

b. Richard paid $4.00 for 8 drinks. How much did each drink cost?

____

c. Daniel had $10.00. After he spent $6.50 on Pip, how much did he have left?

____

d. Find the total amount: $3.50, $1.75, and $2.20.

____

Name **Date**

1.

a. Estimate how many groups of 8 apples there are in this group. Count the apples and write a number sentence to find out how many groups of 8 apples. ___________

b. If Beryl ate 3 apples and Geoff ate 4 apples, how many apples would be left? _____

c. If each apple costs 10¢, how much would it cost me to buy all of the apples? ______

2. Complete these money problems.

a. $2.15 + $1.45

b. $5.00 − $3.50

c. $2.35 x 6

d. 2)$6.66

e. $4.67 x 3

f. $4.85 − $2.62

g. $8.65 + $5.05

h. $9.32 + $4.66

i. 3)$7.71

j. 2)$8.60

k. $5.15 + $6.92

l. $9.05 − $2.55

3. Add the signs to make each number sentence true.

a. 20 ☐ 5 ☐ 15

b. 30 ☐ 6 ☐ 5

c. 27 ☐ 5 ☐ 22

d. 8 ☐ 7 ☐ 56

e. 90 ☐ 9 ☐ 10

f. 42 ☐ 6 ☐ 7

g. 60 ☐ 30 ☐ 30

h. 45 ☐ 5 ☐ 9

i. 23 ☐ 23 ☐ 46

Name ________________________ **Date** ____________

1. Complete the number problems.

a. (16 + 6) – (12 + 3) = ______

b. 30 – (8 + 9) = ______

c. (10 + 8) – (7 + 2) = ______

d. (6 x 7) + (5 x 2) = ______

e. (4 ÷ 2) + (15 ÷ 3) = ______

f. (25 ÷ 5) x (4 + 4) = ______

g. 36 – (7 x 3) = ______

h. (55 + 5) + (6 x 7) = ______

i. 70 – (42 ÷ 6) = ______

j. (6 x 6) + (6 x 6) = ______

k. (20 + 7) – (15 + 2) = ______

l. (100 – 40) ÷ 6 = ______

2. Answer the following money problems. Show all your work. The first one has been done for you.

lollipop
15¢ each

cake
10¢ each

hamburger
$1.25 each

drink
20¢ each

	Name	Had	Bought	How much money spent?	Change
a.	Jarrod	$2.00	1 hamburger and 1 drink	$1.25 + 20¢ = $1.45	$2.00 – $1.45 = 55¢
b.	Brittany	$2.00	1 of everything		
c.	Casey	40¢	2 lollipops		
d.	Ruby	$2.20	1 cake and 1 hamburger		
e.	Bradley	80¢	2 drinks		
f.	Michael	20¢	1 cake		
g.	Connor	60¢	3 lollipops		

3. List the factors of these numbers.

a. 45 ______, ______, ______, ______, ______, and ______

b. 28 ______, ______, ______, ______, ______, and ______

c. 8 ______, ______, ______, and ______

d. 50 ______, ______, ______, ______, ______, and ______

4. Find the answer inside the caterpillar's tail.

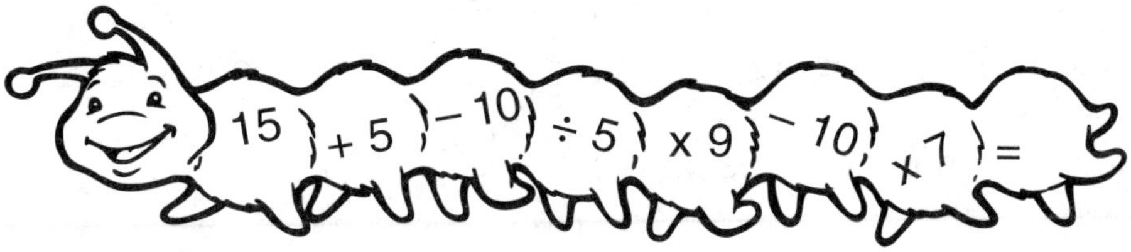

　

Name **Date**

1. Write < or > to make each statement true.

a. 52 + 16 ☐ 648 + 20 b. 50 – 9 ☐ 45 – 3

c. 60 ÷ 10 ☐ 56 ÷ 8 d. 6 x 8 ☐ 5 x 10

e. 32 + 15 ☐ 42 – 2 f. 4 x 4 ☐ 32 – 18

g. 25 ÷ 5 ☐ 56 – 49 h. 7 x 8 ☐ 80 – 32

< is less than
> is more than

2. Complete these magic squares. Be careful to check the operation.

a. (+)

14		30
1		
	31	

b. (−)

65	14	
	7	
42		

c. (x)

	3	6
	3	
8		

3. Find the perimeter of these three shapes.

a.
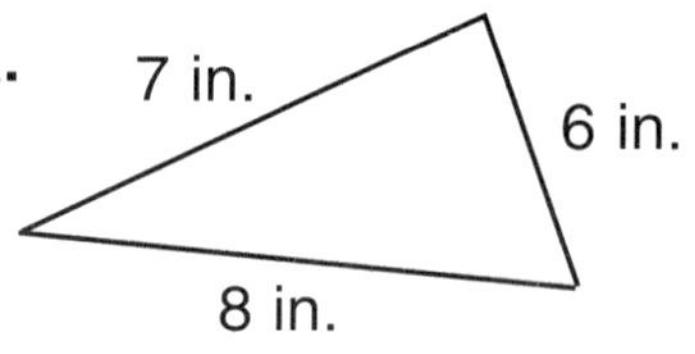

Perimeter = ________

b.
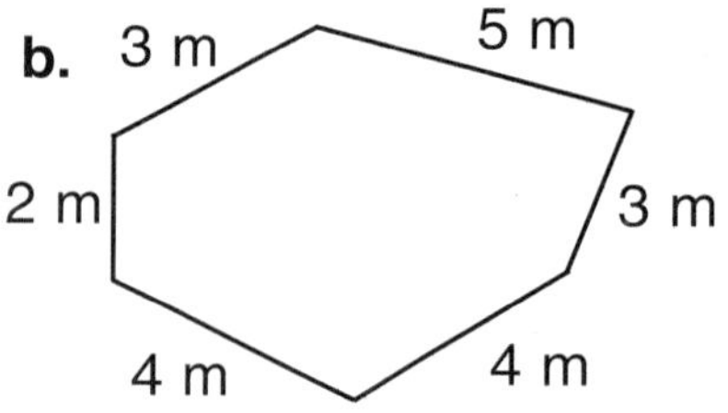

Perimeter = ________

c.
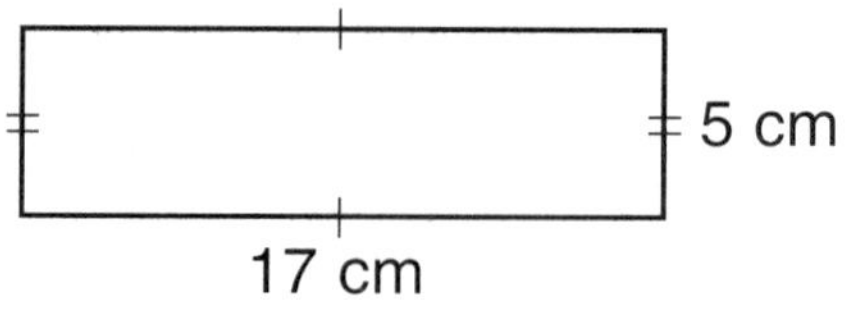

Perimeter = ________

d. Shade the shape with the largest perimeter red.

4. Find the total area of these three shapes.

a.
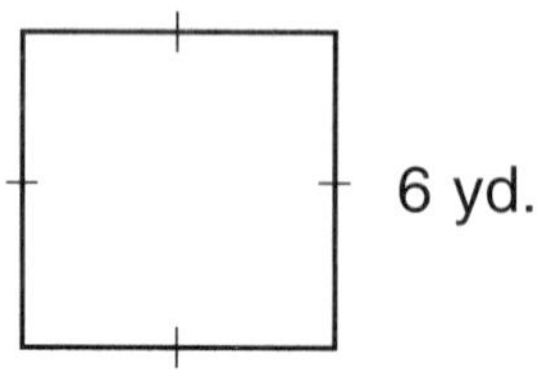

Area = ________ yd.^{2}

b.
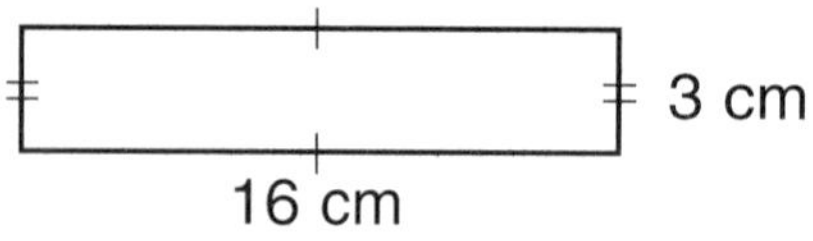

Area = ________ yd.^{2}

c.
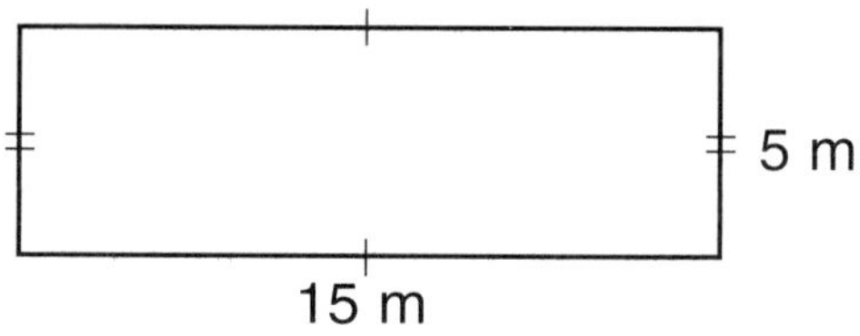

Area = ________ yd.^{2}

d. Shade the shape with the smallest area yellow.

Name **Date**

1. Complete the number problems. Show all work.

a. $\begin{array}{r} 302 \\ +279 \\ \hline \end{array}$ **b.** $\begin{array}{r} 653 \\ -249 \\ \hline \end{array}$ **c.** $\begin{array}{r} 734 \\ \times \quad 6 \\ \hline \end{array}$ **d.** $8\overline{)640}$

e. $6\overline{)426}$ **f.** $\begin{array}{r} 739 \\ \times \quad 2 \\ \hline \end{array}$ **g.** $\begin{array}{r} 666 \\ +\ 532 \\ \hline \end{array}$ **h.** $\begin{array}{r} 701 \\ -323 \\ \hline \end{array}$

i. $\begin{array}{r} 983 \\ +479 \\ \hline \end{array}$ **j.** $\begin{array}{r} 650 \\ -389 \\ \hline \end{array}$ **k.** $\begin{array}{r} 424 \\ \times \quad 4 \\ \hline \end{array}$ **l.** $5\overline{)505}$

2. Read the problems and show all your work.

a. Scott bought 2 cans of soft drinks at 45 cents each. Jillian bought 2 sandwiches at $1.20 each and 2 apples at 30¢ each. How much did their lunches cost together?

(2 x ________) + (2 x ________) + (2 x ________) = ________

b. Jane sent Matthew to the shop to buy 4 items all costing the same amount. She gave him $1.00 and received no change. How much did each item cost?

$4\overline{)\$1.00}$

c. Rose collected 15 shells, Tilly collected 32 shells, and Thomas collected 26 shells. How many did they take home to add to their collection?

____

d. The fruit seller had 2 dozen apples in the basket. Five were bad and had to be thrown away. How many were left in the basket?

____

3. **a.** 600 + 70 + 4 = ______ **b.** 470 + 62 + 3 = ______ **c.** 220 + 63 + 7 = ______

d. 530 + 40 + 9 = ______ **e.** 330 + 54 + 7 = ______ **f.** 190 + 55 + 6 = ______

Name	**Date**

1. Show all work.

a. $\$2.05 + \$3.60$

b. $732 - 198$

c. $6\overline{)366}$

d. 432×6

e. $2\overline{)312}$

f. 637×5

g. $732 - 498$

h. $983 + 654$

2. Complete.

a. (6 + 6) + (7 x 3) = ________

b. 730 + 40 + 3 = ________

c. (80 ÷ 8) – (2 x 3) = ________

d. 800 + 70 + 6 = ________

e. 52 ÷ 6 = ________

f. (9 x 8) – 10 = ________

3. Solve these problems.

a. Walter had to ride 161 miles during the first week of competition. How far did he have to ride each day?

____

b. Cindy bought half a dozen oranges at 30¢ each and 4 bananas at 20¢ each. How much change did she receive from $5.00?

____

c. Lou worked 13 hours on Monday, 12 hours on Tuesday, 15 hours on Wednesday, 10 hours on Thursday, and 5 hours on Friday. How many hours did he work that week?

____

4. Find the perimeter and area of these three shapes.

a.

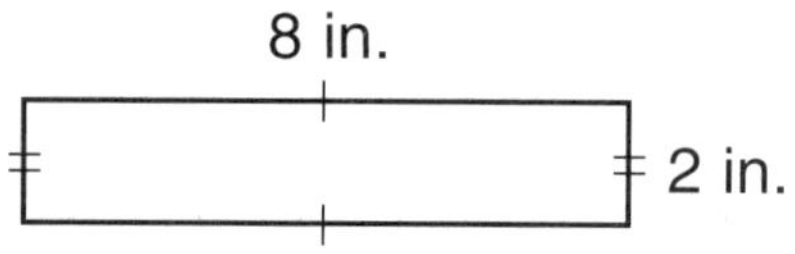

Perimeter = ________
Area = ________

b.

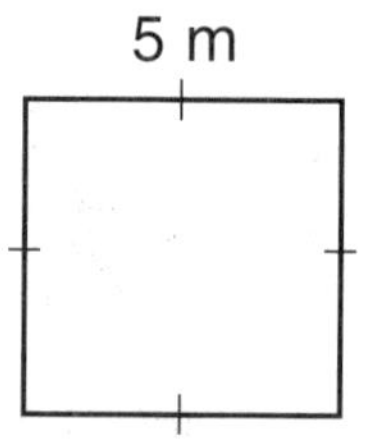

Perimeter = ________
Area = ________

c.

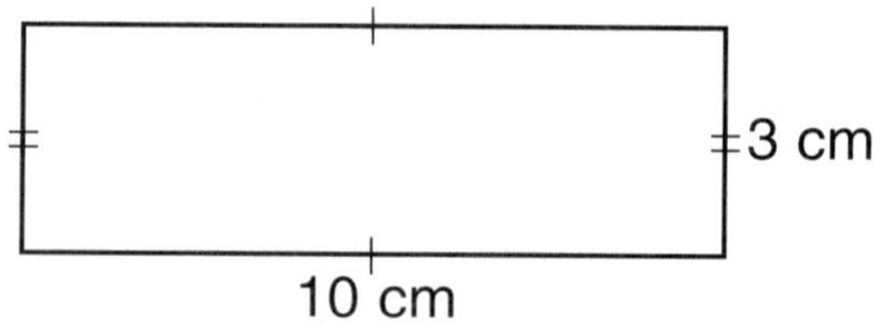

Perimeter = ________
Area = ________

5. Follow the arrows and signs to the final answers.

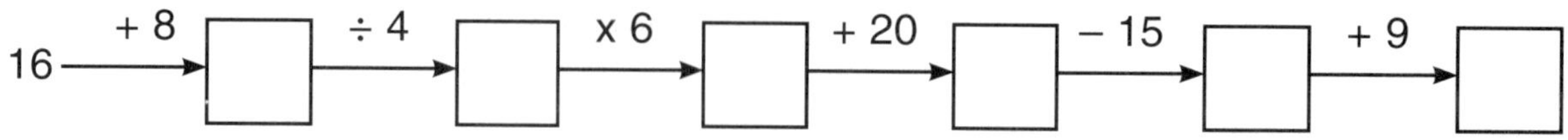

MORE NUMBER FACTS PRACTICE

Unit 2

Patterns
Brackets
Money
Sum, product
Problem solving

Objectives

- *use number skills involving whole numbers to solve problems*
- *use one or more strategies to solve mathematical problems*
- *check whether answers to problems are correct and sensible*
- *model and explain patterns related to number facts for the four operations*
- *use a variety of mental, written, and informal strategies to approximate, calculate, and represent solutions to multiplication, division, addition, and subtraction problems*
- *comprehend and solve written word problems*

Language

subtraction, addition, division, multiplication, groups, checking, operation, signs, parentheses, greater than, less than, patterns, factors, solve, decrease, increase, sum, product, true, statement, spent, change

Materials/Resources

coins (optional), blank paper, dice (optional)

Contents of Student Pages

* *Materials needed for each reproducible student page*

Remember

Before starting ensure each student:

- ❑ *understands the importance of showing all work neatly and correctly.*
- ❑ *develops the habit of checking.*
- ❑ *is reminded to watch for the various operation signs.*

Additional Activities

- Students make own magic squares for friends to solve.
- Flash card games where the "audience" calls out the operation.
- Play a number game. Two players have three dice each. When they roll them, the first to add all three numbers wins. This can work with all operations with various numbers of dice.

Answers

Page 85 Mixed Number Facts

1. a. 1,085 e. 547 i. 163 r1
   b. 275 f. 3,365 j. 137
   c. 4,473 g. 18 k. 4,641
   d. 70 r2 h. 5,072 l. 442
2. a. April 1 b. 240 seats c. $175
3. a. 50 e. 44 i. 30
   b. 39 f. 61 j. 88
   c. 33 g. 12 k. 52
   d. 95¢ h. 42 l. 4
4. 100, 10, 70, 35, 55, 11, 20

Page 86 Number Wheels and More

1. a.

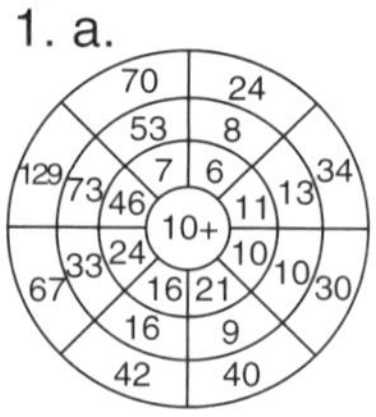

b.

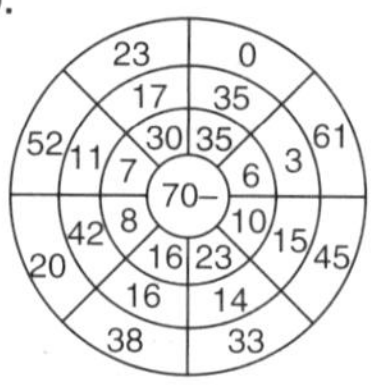

c.

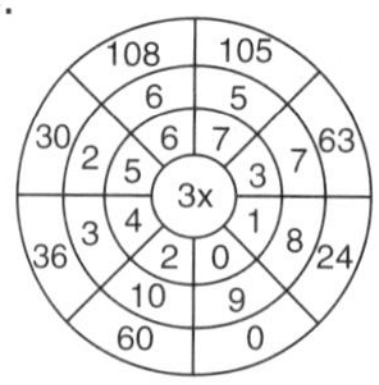

2. a. – e. – i. ÷, –
   b. ÷ f. x, – j. ÷
   c. +, – g. + k. x
   d. x h. + l. ÷, x
3. a. 44, 51, 58 d. 33, 55, 77 g. 48, 8
   b. 80, 100, 120 e. 90, 150, 180 h. 200, 400, 600
   c. 62, 53, 44 f. 52, 32, 12
4. a. 127 r1 e. 114 r1 i. 182
   b. 102 r2 f. 166 r1 j. 169 r3
   c. 113 r2 g. 61 r1 k. 129 r3
   d. 90 h. 71 r3 l. 150

Page 87 Missing Numbers

1. a.

x	8	6	10	12	9
6	48	36	60	72	54
9	72	54	90	108	81

b.

+	26	23	31	19	7
10	36	33	41	29	17
30	56	53	61	49	37

2. a. 36 e. 80 i. 11
   b. 64 f. 50 j. 30
   c. 30 g. 74
   d. 62 h. 27
3. a. $24.45 e. $1.30 i. $106.35
   b. $6.25 f. $40.85 j. $20.24
   c. $54.00 g. $23.55 k. $15.12
   d. $1.08 h. $28.75 l. $7.00
4. a. 48 e. 20 i. 92
   b. 38 f. 28 j. 270
   c. 52 g. 32
   d. 102 h. 0

Page 88 Crossnumber and More

1. Teacher to check individual word problems.
   a. $14.18 c. $47.70
   b. $3.70 d. $2.85
2. a. 60, 6, 42, 60, 30, 90, 45
   b. 20, 100, 50, 5, 30, 15, 51
3.

[1] 4	4		[2] 6	8		[3] 2
5		[4] 4	8			7
	[5] 4	9			[6] 1	5
[7] 8	0		[8] 1		9	
5			[9] 5	7	8	
	[10] 3	0	9			[11] 9

4. a. +, +, + c. ÷, +, ÷ e. x, –, +
   b. x, +, x d. +, –, + f. +, x, x

Page 89 Sum and Product

1. a. 16, 63 c. 45, 324 e. 99, 558
   b. 17, 60 d. 64, 448
2. a. 180 e. 143 i. 30 r4
   b. 211 f. 5,341 j. 2,583
   c. 330 g. 333 k. 135
   d. 238 h. 901 l. 1,021
3. a. $60.00 b. $850.00 c. 137 r4

Page 90 Assessment

1. a. 1,530 d. 122 g. 1,966
   b. 516 e. 243 h. 53 r5
   c. 3,068 f. 153
2. a. 41 e. 82 h. 125
   b. 22 f. 112 i. 51
   c. 35 g. 72 j. 19
   d. 10
3. 70, 7, 42, 50, 20, 100, 55, 11
4. a. ÷ d. +, + g. x
   b. + e. –, + h. –
   c. – f. x, – i. ÷
5. a. 2 children
   b. $77.50

Name **Date**

1. Complete the number problems.

a. $\begin{array}{r} 736 \\ +\ 349 \\ \hline \end{array}$ **b.** $\begin{array}{r} 600 \\ -\ 325 \\ \hline \end{array}$ **c.** $\begin{array}{r} 639 \\ \times\ \ \ 7 \\ \hline \end{array}$ **d.** $7\overline{)492}$

e. $\begin{array}{r} 901 \\ -\ 354 \\ \hline \end{array}$ **f.** $\begin{array}{r} 673 \\ \times\ \ \ 5 \\ \hline \end{array}$ **g.** $\begin{array}{r} 207 \\ -\ 189 \\ \hline \end{array}$ **h.** $\begin{array}{r} 634 \\ \times\ \ \ 8 \\ \hline \end{array}$

i. $2\overline{)327}$ **j.** $5\overline{)685}$ **k.** $\begin{array}{r} 663 \\ \times\ \ \ 7 \\ \hline \end{array}$ **l.** $\begin{array}{r} 729 \\ -\ 287 \\ \hline \end{array}$

2. Read these problems and solve them. Show all work.

a. A rock concert arranged for the 11th of March was postponed for three weeks. On what date was it held?

____

b. Rows of seats at the concert are labeled A to L. If there are 20 seats in each row, how many seats are there altogether?

____

c. The Neale family is all going to the concert. They have to purchase 5 tickets at $35.00 each. How much did it cost to take the family to the concert?

____

3.
a. Increase 35 by 15 ________
b. The sum of 13 and 26 ________
c. Add 17 and 16 ________
d. 20 cents plus 75 cents ________
e. Decrease 60 by 16 ________
f. 75 take away 14 ________
g. The product of 6 and 2 is________
h. Increase 25 by 17 ________
i. Decrease 41 by 11 ________
j. Multiply 8 by 11 ________
k. Subtract 6 from 58 ________
l. Divide 36 by 9 ________

4. What is the final answer?

45 —(+ 55)→ ☐ —(÷ 10)→ ☐ —(x 7)→ ☐ —(– 35)→ ☐ —(+ 20)→ ☐ —(÷ 5)→ ☐ —(+ 9)→ ◯

Name **Date**

1. Complete each number wheel. Check the signs and the double operations.

a.

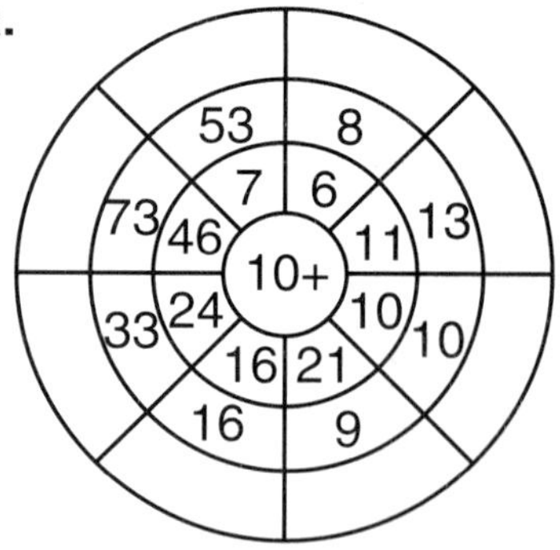

b.

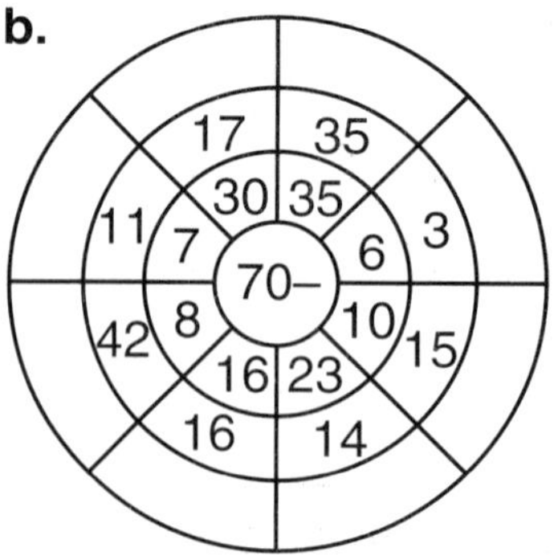

c.

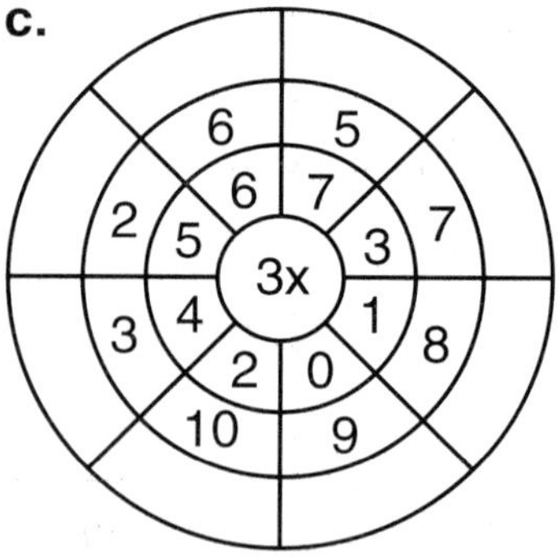

2. Write an operation sign (+, –, ÷, x) in the box to make the number sentence true.

a. 50 ☐ 10 = 40 **b.** 56 ☐ 8 = 7 **c.** 30 ☐ 50 = 90 ☐ 10

d. 4 ☐ 15 = 60 **e.** 63 ☐ 17 = 46 **f.** 5 ☐ 5 = 30 ☐ 5

g. 65 ☐ 20 = 85 **h.** 44 ☐ 35 = 79 **i.** 63 ☐ 3 = 28 ☐ 7

j. 77 ☐ 7 = 11 **k.** 10 ☐ 9 = 90 **l.** 64 ☐ 8 = 4 ☐ 2

3. Complete the patterns.

a. 23, 30, 37, ______ , ______ , ______

b. 20, 40, 60, ______ , ______ , ______

c. 89, 80, 71, ______ , ______ , ______

d. 11, 22, ______ , 44, ______ , 66, ______

e. 30, 60, ______ , 120, ______ , ______

f. 62, ______ , 42, ______ , 22, ______

g. 88, 68, ______ , 28, ______

h. 100, ______ , 300, ______ , 500, ______

4. Write the answer and remainders to these division problems.

a. 6)763 **b.** 4)410 **c.** 5)567 **d.** 7)630

e. 8)913 **f.** 2)333 **g.** 5)306 **h.** 4)287

i. 2)364 **j.** 4)679 **k.** 6)777 **l.** 3)450

Name **Date**

1. Find the rule for each table to complete it.

a.

	8		10		9
6		36	60	72	
					81

b.

	26	23		19	7
10	36			29	
30			61		

2. Complete the following using the parentheses.

a. (13 + 46) – (20 + 3) = ________

b. (63 – 31) x 2 = ________

c. (53 + 10) – (17 + 16) = ________

d. (70 – 50) + (6 x 7) = ________

e. (80 ÷ 8) + (10 x 7) = ________

f. (6 x 7) + (64 ÷ 8) = ________

g. (70 ÷ 7) + (8 x 8) = ________

h. (9 x 7) – (6 x 6) = ________

i. (7 x 11) – (6 x 11) = ________

j. (13 – 7) x (40 ÷ 8) = ________

3. Solve these money problems.

a. \$19.65 + \$4.80

b. \$20.00 – \$13.75

c. \$6.75 x 8

d. 8) \$8.64

e. 6) \$7.80

f. \$13.99 + \$26.86

g. \$30.50 – \$6.95

h. \$5.75 x 5

i. \$62.85 + \$43.50

j. \$29.16 – \$8.92

k. \$3.78 x 4

l. 7) \$49.00

4. How many legs have:

a. 6 spiders? ________

b. 19 birds? ________

c. 13 horses? ________

d. 17 ants? ________

e. 10 ducks? ________

f. 14 boys? ________

g. 16 girls? ________

h. 30 fish? ________

i. 23 dogs? ________

j. 45 insects? ________

Name **Date**

1. Write a problem of your own to match each of the number problems below. Work out the answer.

a.
$6.75
$5.43
+ $2.00

b.
$12.65
− $8.95

c.
$7.95
x 6

d.
6)$17.10

2. Follow the arrows to find the answer.

a. 52 —+ 8→ ☐ —÷ 10→ ☐ —x 7→ ☐ —+ 18→ ☐ —÷ 2→ ☐ —x 3→ ☐ —− 45→ ◇

b. 18 —+ 2→ ☐ —x 5→ ☐ —− 50→ ☐ —÷ 10→ ☐ —x 6→ ☐ —− 15→ ☐ —+ 36→ ◇

3. Complete the crossnumber puzzle.

Down
1. 36 + 9
2. 100 − 32
3. 536 − 261
4. 98 ÷ 2
5. 90 − 50
6. 200 − 2
7. 17 x 5
8. 53 x 3
11. 99 ÷ 11

Across
1. 4 x 11
2. 15 + 53
4. 58 − 10
5. 7 x 7
6. 45 ÷ 3
7. 36 + 44
9. 500 + 76 + 2
10. 864 − 555

4. Add the appropriate signs to make each statement true.

a. (5 ☐ 5) ☐ (5 ☐ 5) = 20

b. (5 ☐ 5) ☐ (5 ☐ 5) = 50

c. (5 ☐ 5) ☐ (5 ☐ 5) = 2

d. (5 ☐ 5) ☐ (5 ☐ 5) = 0

e. (5 ☐ 5) ☐ (5 ☐ 5) = 15

f. (5 ☐ 5) ☐ (5 ☐ 5) = 250

Name **Date**

1.

	Numbers	Sum	Product
a.	7 and 9		
b.	12 and 5		
c.	36 and 9		
d.	8 and 56		
e.	6 and 93		

Sum → add

Product → multiply

2.

a. 73 + 65 + 27 + 15

b. 93 + 67 + 32 + 19

c. 66 + 77 + 88 + 99

d. 53 + 42 + 50 + 93

e. 632 − 489

f. 763 x 7

g. 802 − 469

h. 9)8109

i. 7)214

j. 369 x 7

k. 432 − 297

l. 489 + 532

3. Read and solve these problems.

a. The Smith family paid $42.00 to see the concert in the park. There are four children in the family and each child received $4.50 to spend on something to eat and drink. How much did it cost the Smith's for the entire night?

____

b. James was saving to go overseas. He had saved $850 for the flight and $300 for spending money. How much more did he have to save before reaching the $2,000 amount?

____

c. Jane had 963 beads to thread onto 7 necklace bands. How many beads would be on each necklace? Would there be any left over?

____

Name **Date**

1. Solve these number problems. Show all work and check carefully.

a. $\begin{array}{r} 637 \\ +893 \\ \hline \end{array}$

b. $\begin{array}{r} 702 \\ -186 \\ \hline \end{array}$

c. $\begin{array}{r} 767 \\ \times \quad 4 \\ \hline \end{array}$

d. $6\overline{)732}$

e. $\begin{array}{r} 63 \\ 79 \\ 84 \\ +17 \\ \hline \end{array}$

f. $\begin{array}{r} 632 \\ -479 \\ \hline \end{array}$

g. $\begin{array}{r} 983 \\ \times \quad 2 \\ \hline \end{array}$

h. $7\overline{)376}$

2. Read the problems carefully, then solve them.

a. Increase 25 by 16 ______

b. Decrease 39 by 17 ______

c. The sum of 17 and 18 ______

d. Divide 70 by 7 ______

e. 97 take away 15 ______

f. The product of 16 and 7 ______

g. Multiply 8 and 9 ______

h. Add 53 and 72 ______

i. Subtract 51 from 102 ______

j. Difference between 35 and 16 ______

3. What is the final answer?

63 —+ 7→ ☐ —÷ 10→ ☐ —x 6→ ☐ —+ 8→ ☐ —– 30→ ☐ —x 5→ ☐ —– 45→ ☐ —÷ 5→ ◇

4. Write the correct sign (+, –, x, or ÷) so that the statement is true.

a. 125 ☐ 5 = 25

b. 36 ☐ 42 = 78

c. 80 ☐ 32 = 48

d. 35 ☐ 35 = 60 ☐ 10

e. 85 ☐ 16 = 60 ☐ 9

f. 17 ☐ 2 = 40 ☐ 6

g. 4 ☐ 12 = 48

h. 89 ☐ 40 = 49

i. 93 ☐ 3 = 31

5. Show all work on a separate sheet of paper.

a. On Wednesday there were 16 boys and 17 girls in class. If there are usually 35 children in the class, how many were absent?

____

b. Mrs. G took some of her class to the zoo. It cost $7.50 for each student and $10 for the teacher. How much did the teacher pay if she took 9 children and herself to the zoo?

____

NUMBER PATTERNS

Through work with number patterns students learn that mathematics does not consist entirely of straight operational work with numbers and shapes. Skills of observation, generalization, visualization, and approximation must be developed. In these units, number patterns are explored through sequencing, following rules to create patterns, completing and creating shape patterns, number grids, and finding and explaining the relationships for given patterns. The activity page has fun patterns for students to explore. There are two assessment pages.

NUMBER PATTERN PRACTICE

Unit 1

Number patterns
Shape patterns
Grids
Making patterns

Objectives

- *appreciate the importance of visualization when solving problems*
- *predict or approximate the value of a given term in a sequence*
- *check whether answers are correct and sensible*
- *appreciate that mathematics involves observing, generalizing, and representing patterns and relationships*
- *generate and describe number patterns using a variety of strategies*

Language

pattern, sequence, shapes, puzzles

Materials/Resources

colored pencils, blank paper, counters

Contents of Student Pages

** Materials needed for each reproducible student page*

Remember

Before starting ensure each student:

- ❑ *reads questions twice.*
- ❑ *develops the habit of checking.*
- ❑ *is encouraged to ask for help when needed.*

Additional Activities

- ❑ *Students cut out various 2-D shapes and place them in a pattern. A partner guesses what is next.*
- ❑ *Patterns and shapes are made using blocks.*
- ❑ *Students draw a detailed pattern in a shape. A partner has to copy the design. This will develop checking skills.*

Answers

Page 94 Mixed Patterns

1. a. 8, 10, 12, 18, 20, 22
   b. 21, 24, 27, 36, 39, 42
   c. 30, 40, 50, 60, 80, 90
   d. 52, 47, 37, 32, 27, 17
   e. 35, 23, 15, 11, 3
   f. 25, 30, 35, 40, 50, 55
2. a. 11, 16, 21, 26, 31
   b. 18, 28, 38, 48, 58
   c. 36, 29, 22, 15, 8
   d. 53, 50, 47, 44, 41
3. a.
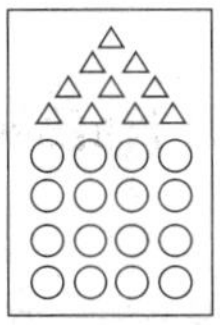
   b.

Page 95 Drawing Patterns

Teacher to check individual work.

Page 96 Making Patterns

Teacher to check individual work.

Page 97 Grids

1. a.

	5	6	11	9	3
x6	30	36	66	54	18
+3	8	9	14	12	6

b.

	2	5	6	8	9
+10	12	15	16	18	19
x10	20	50	60	80	90

c.

	10	7	3	11	5
+8	18	15	11	19	13
x7	70	49	21	77	35

d.

	6	9	11	2	20
+12	18	21	23	14	32
–2	4	7	9	0	18

e.

	10	8	2	0	6
x4	40	32	8	0	24
+15	25	23	17	15	21

f.

	20	9	15	12	17
–6	14	3	9	6	11
+20	40	29	35	32	37

2. a. Spider—16, 24, 32, 40
   b. Dog—8, 12, 16, 20
   c. Birds—8, 18, 14, 24

Page 98 More Mixed Patterns

1. a. 3, 16, 19, 16, 19, 3, 19, 16
   b. 40, 35, 40, 16, 40, 16, 35, 40, 16
   c. 2, 1, 2, 2, 2, 2, 1, 2
   d. 7, 8, 6, 8, 7, 8, 6, 8, 6
   e. 19, 29, 9, 19, 39, 19, 29, 39
2. a. circle, rectangle, triangle, circle, circle, rectangle, triangle, rectangle
   b. arrow up, arrow left, arrow right, arrow left, arrow right, arrow up, arrow right, arrow up
   c. heart, square, circle, square, heart, heart, square
   d. circle, circle, circle, circle, circle, triangle, circle, circle, triangle
   e. horizontal rectangle, vertical rectangle, horizontal rectangle, vertical rectangle, horizontal rectangle, horizontal rectangle, vertical rectangle, vertical rectangle, horizontal rectangle
3. a. 3, 6, 9, 18, 24, 33
   b. 2, 4, 6, 14, 18, 24
   c. 2, 4, 6, 10, 12, 18
4. a. 11, 17, 23, 29
   b. 25, 40, 55, 70
   c. 32, 28, 24, 20
   d. 9, 27, 81, 243
   e. 52, 42, 32, 22
   f. 20, 80, 320, 1,280
   g. 24, 33, 42, 51
   h. 45, 36, 27, 18

Page 99 Hundred Squares

Teacher to check individual work.

Page 100 Assessment

1. a. 24, 48, 96
   b. 25, 30, 35
   c. 20, 18, 16
   d. 28, 35, 42
   e. 32, 42, 52
   f. 17, 23, 30
2. a.

	6	2	10	3	9
x4	24	8	40	12	36
+6	12	8	16	9	15

b.

	12	9	5	3	6
–3	9	6	2	0	3
x5	60	45	25	15	30

3. Teacher to check.
4. a. triangle, square, triangle, square
   b. diamond, wavy line, diamond, wavy line
   c. circle, circle, horizontal rectangle, circle
   d. horizontal rectangle, circle, circle, horizontal rectangle
   e. circle, triangle, square, horizontal rectangle
   f. square, square, square, triangle
5. 4, 7, 10, 16, 22, 31

Page 101 Activity—Game and Patterns

1. a. ∂ 7 ʕ 8
   b.

   c. 12:55, 1:00, 1:05
   d.

2.

R	B	G	Y
Y	G	B	R
B	R	Y	G
G	Y	R	B

Name **Date**

1. Complete these patterns.

   a. 2, 4, 6, ______ , ______ , ______ , 14, 16, ______ , ______ , ______

   b. 15, 18, ______ , ______ , ______ , 30, 33, ______ , ______ , ______

   c. 10, 20, ______ , ______ , ______ , ______ , 70, ______ , ______ , 100

   d. 62, 57, ______ , ______ , 42, ______ , ______ , ______ , 22, ______

   e. 39, ______ , 31, 27, ______ , 19, ______ , ______ , 7, ______

   f. 15, 20, ______ , ______ , ______ , ______ , 45, ______ , ______ , 60

2. Use the rule in each square to fill in the missing numbers.

   a. [+5] 6 → ( ) → ( ) → ( ) → ( ) → ( )

   b. [+10] 8 → ( ) → ( ) → ( ) → ( ) → ( )

   c. [−7] 43 → ( ) → ( ) → ( ) → ( ) → ( )

   d. [−3] 56 → ( ) → ( ) → ( ) → ( ) → ( )

3. Draw the next set of shapes in these patterns.

   a.

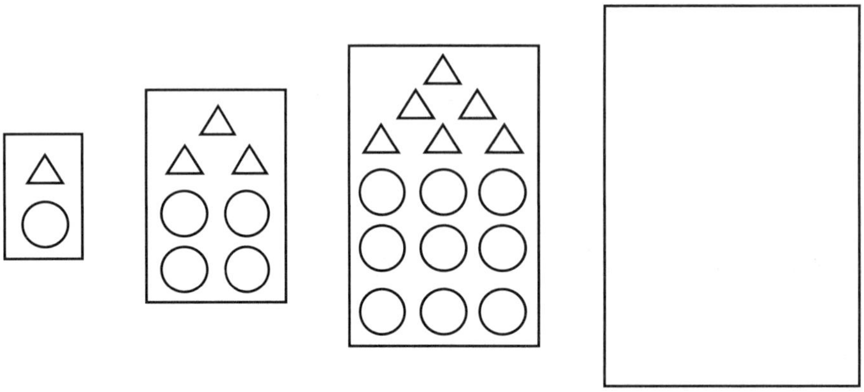

   b.

Name **Date**

1. Complete each of the shapes so that they match the first shape.

a.

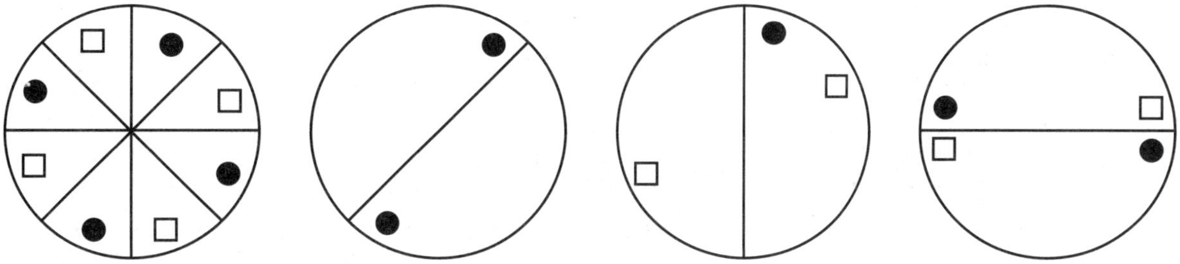

b.

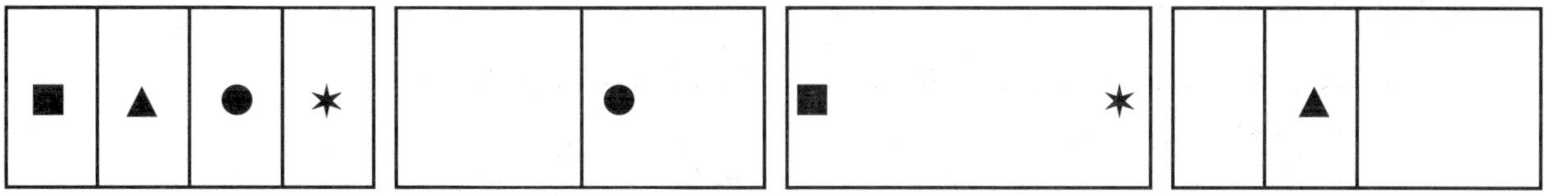

c.

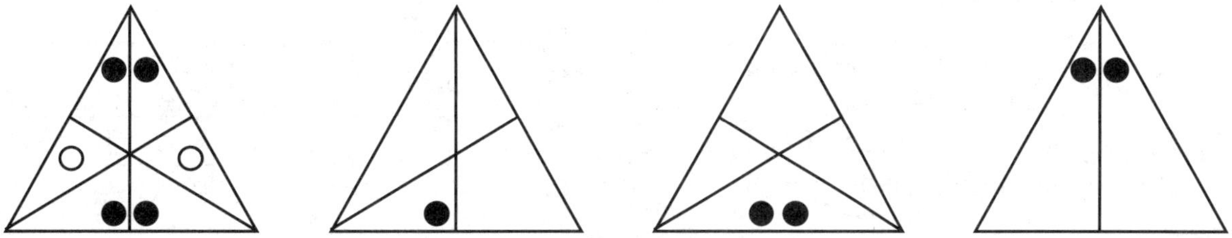

d.

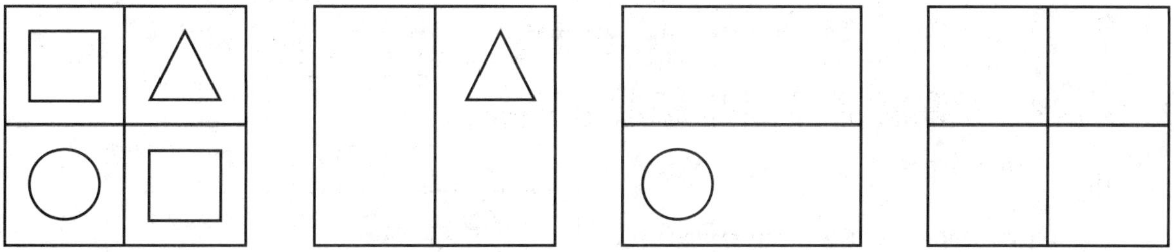

e.

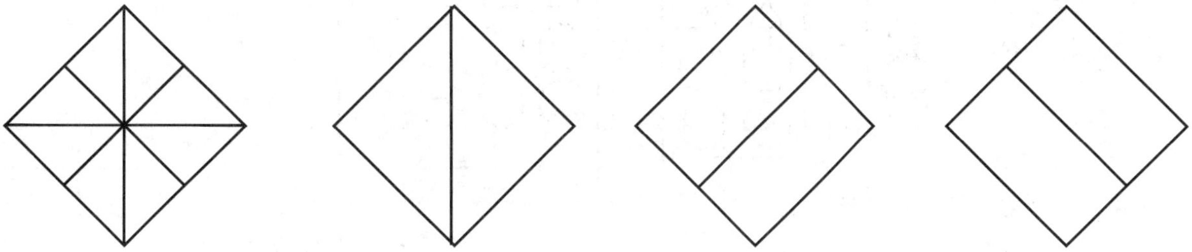

f.

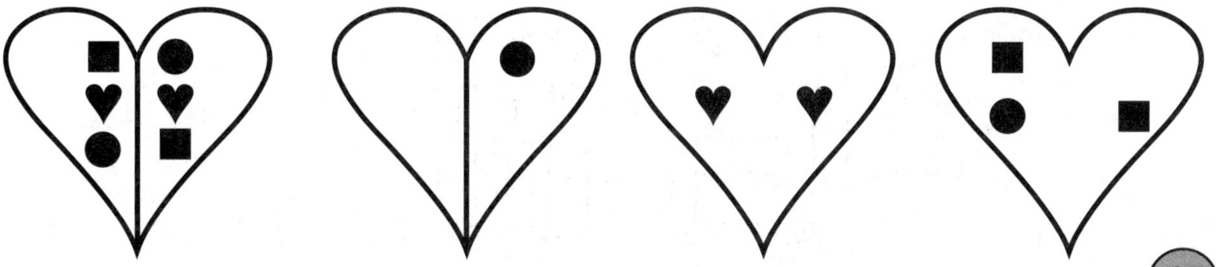

Name **Date**

1. Use five different colors to make a pattern in the boxes below.

2. Follow the steps to create each pattern.

a. Color 2 squares red, 1 square blue.

Color 4 squares red, 3 squares blue.

Color 6 squares red, 5 squares blue.

Color the next set of squares to continue the pattern.

b. Checkmark 2 squares, color 2 squares.

Checkmark 3 squares, color 3 squares.

Checkmark 4 squares, color 4 squares.

Complete the pattern in these boxes.

c. Color 1 square, put dots in 3 squares.

Color 2 squares, put dots in 6 squares.

Color 3 squares, put dots in 9 squares.

Complete the pattern in these boxes.

d. Color 1 square, leave 1 blank.

Color 2 squares, leave 2 squares blank.

Color 3 squares, leave 3 squares blank.

Complete the pattern in these boxes.

Name **Date**

1. Use the rules to complete the patterns of numbers in the tables.

a.

●	5	6	11	9	3
x 6					
+ 3					

b.

♥	2	5	6	8	9
+ 10					
x 10					

c.

▲	10	7	3	11	5
+ 8					
x 7					

d.

⯃	6	9	11	2	20
+ 12					
− 2					

e.

■	10	8	2	0	6
x 4					
+ 15					

f.

☆	20	9	15	12	17
− 6					
+ 20					

2. Complete these patterns.

a.

Spiders	Number of legs
1	8
2	
3	
4	
5	

b.

Dogs	Number of legs
1	4
2	
3	
4	
5	

c.

Birds	Number of legs
1	2
4	
9	
7	
12	

Name **Date**

1. Complete these number patterns.

   a. 3, 19, 16, 3, 19, 16, ___ , 19, ___ , 3, ___ , ___ , 3, ___ , 16, ___ , ___ , ___

   b. 35, 40, 16, 35, ___ , 16, ___ , ___ , ___ , 35, ___ , ___ , ___ , ___ , ___

   c. 1, 2, 2, 1, 2, ___ , ___ , 2, ___ , 1, ___ , ___ , 1, 2, ___ , ___ , 2, ___

   d. 6, 7, 8, 6, ___ , ___ , ___ , 7, ___ , 6, ___ , ___ , ___ , 7, ___ , ___

   e. 9, 19, 29, 39, 9, ___ , ___ , 39, ___ , ___ , 29, ___ , 9, ___ , ___ , ___

2. Complete these picture patterns.

   a. 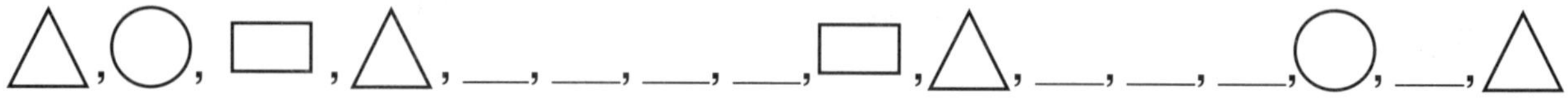

   b. 

   c. 

   d. 

   e. 

3. Look at each pattern and complete the tables.

   a. 

Number of triangles	1	2	3	6	8	11
Number of sides						

   b. 

Number of circles	1	2	3	7	9	12
Number of dots						

   c. 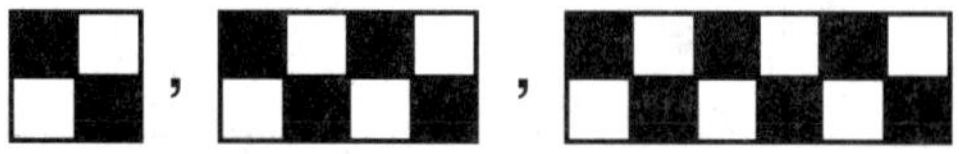

Number of colored squares	2	4	6	10	12	18
Number of uncolored squares						

4. Use the first number and follow the instruction to write the next 4 numbers.

   a. 5; "add 6" ___ , ___ , ___ , ___

   b. 10; "add 15" ___ , ___ , ___ , ___

   c. 36; "take away 4" ___ , ___ , ___ , ___

   d. 3; "multiply by 3" ___ , ___ , ___ , ___

   e. 62; "subtract 10" ___ , ___ , ___ , ___

   f. 5; "multiply by 4" ___ , ___ , ___ , ___

   g. 15; "add 9" ___ , ___ , ___ , ___

   h. 54; "take away 9" ___ , ___ , ___ , ___

Name **Date**

1. Some squares on the grid have been shaded. Finish shading the squares to continue the pattern.

a. 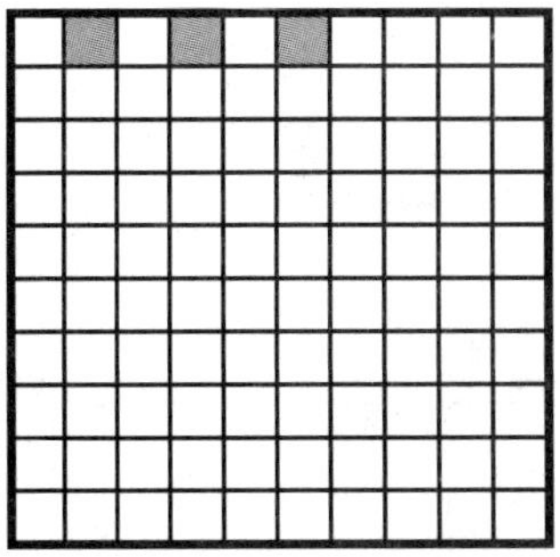

b. 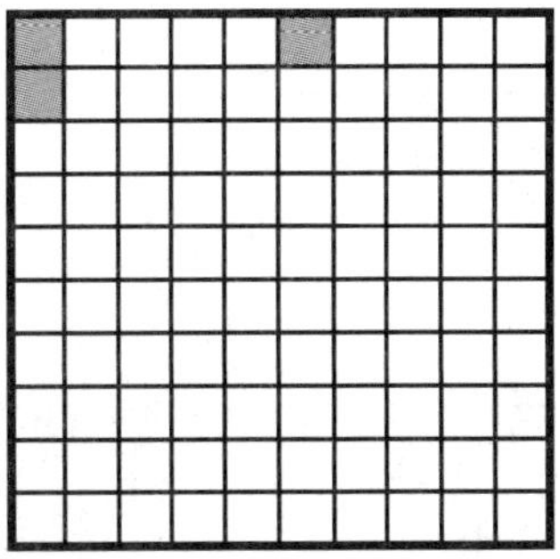

c. 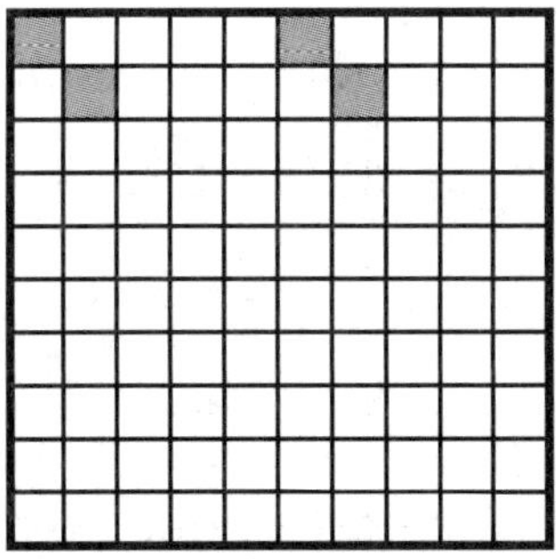

2. Read each rule below the squares and color the squares to make the patterns.

a. 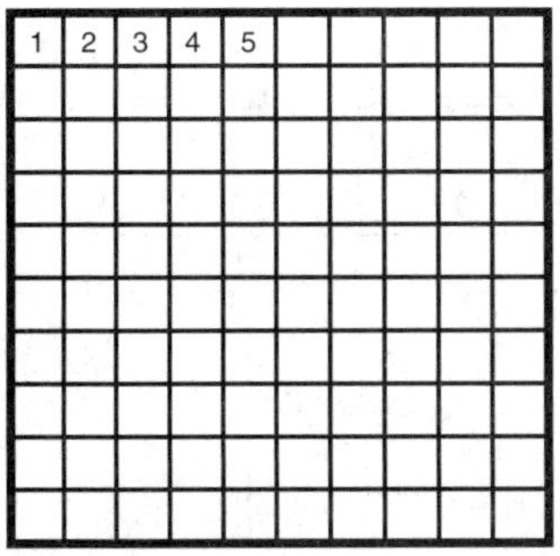

Start at 5 and add 7.

b. 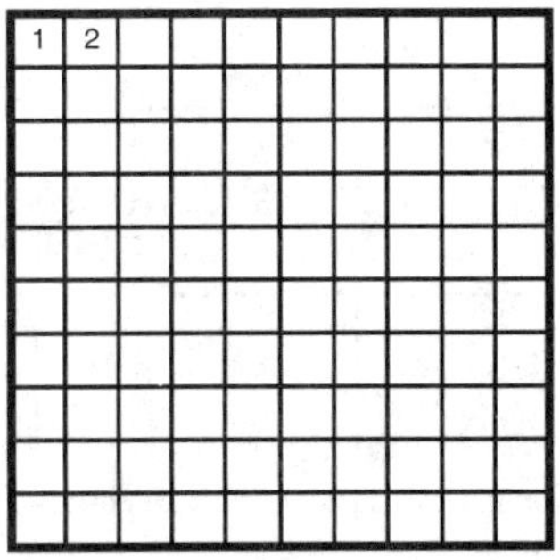

Start at 2 and add 10.

c. 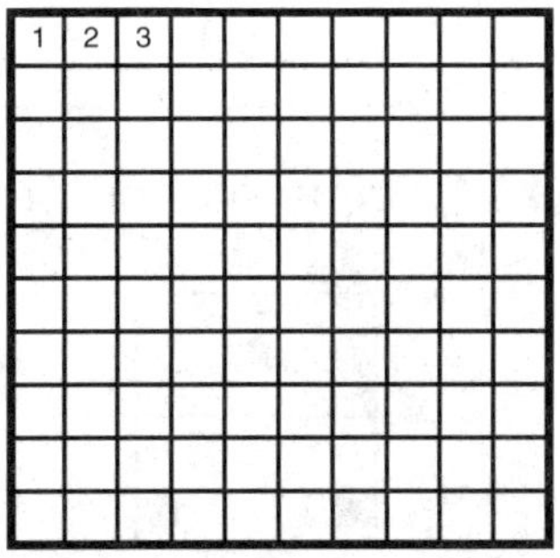

Start at 3 and add 8.

d. 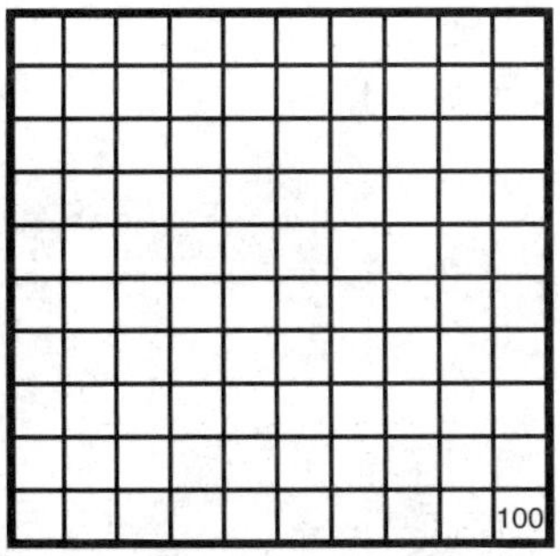

Start at 100 and subtract 6.

e. 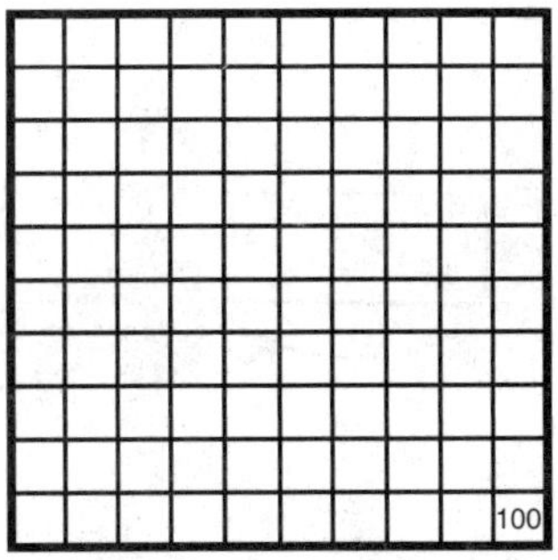

Start at 100 and subtract 12.

f. 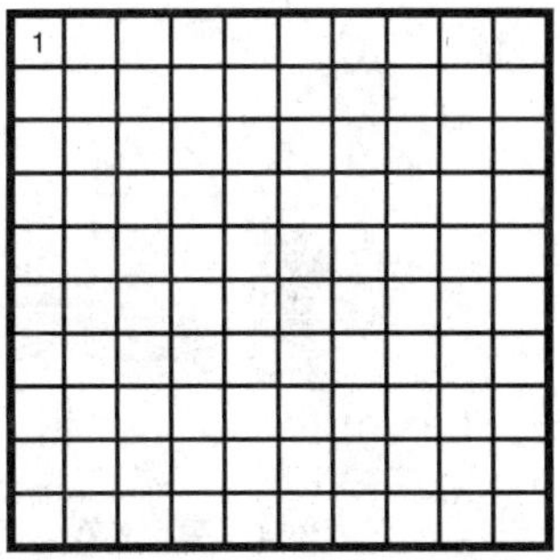

Start at 1 and add 9.

3. Make up your own rule and color the patterns.

a. 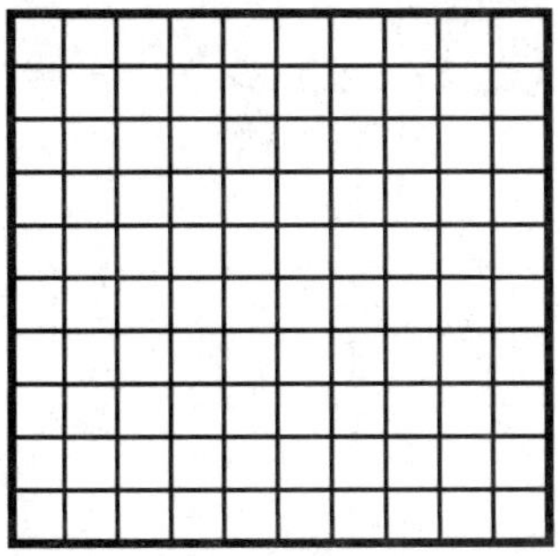

b. 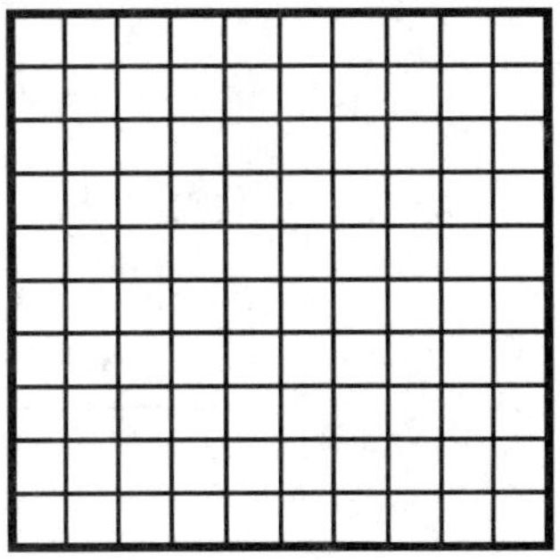

c. 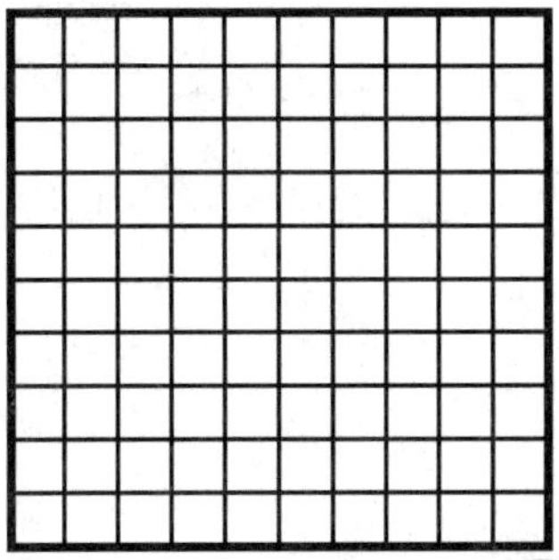

Name **Date**

1. Complete these number patterns.

   a. 3, 6, 12, ______ , ______ , ______

   b. 10, 15, 20, ______ , ______ , ______

   c. 26, 24, 22, ______ , ______ , ______

   d. 7, 14, 21, ______ , ______ , ______

   e. 2, 12, 22, ______ , ______ , ______

   f. 3, 5, 8, 12, ______ , ______ , ______

2. Use the rules to complete each table.

a.

	6	2	10	3	9
x 4					
+ 6					

b.

	12	9	5	3	6
– 3					
x 5					

3. Complete each of the patterns on the shapes to match the first one.

a.

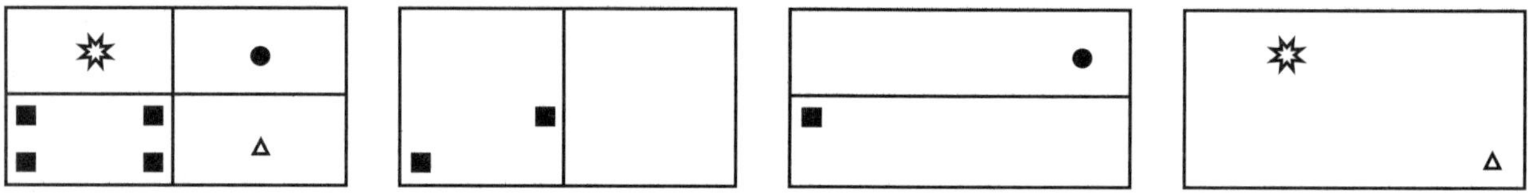

b.

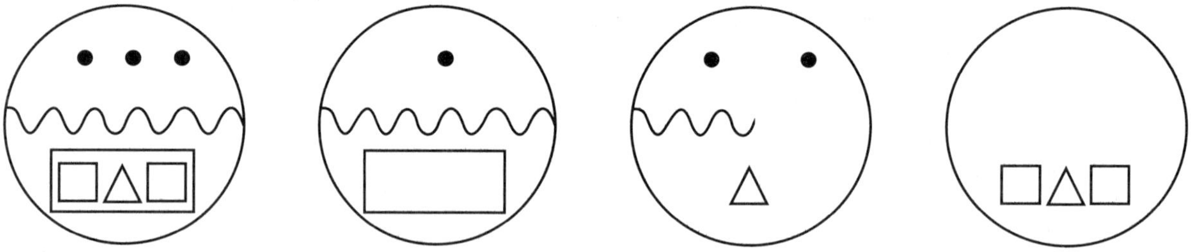

4. Complete the patterns.

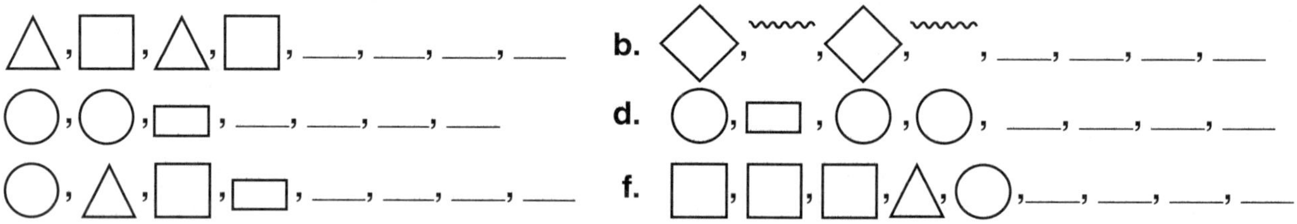

5. Look at each pattern and complete the table.

Number of squares	1	2	3	5	7	10
Number of sides						

Name **Date**

1. Complete these patterns.

a.  ______ ______ ______ ______

b. 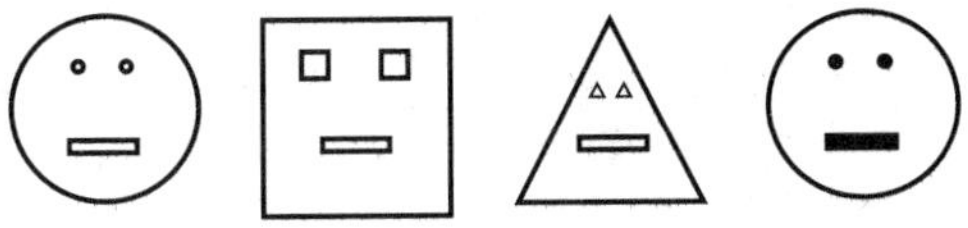 ______ ______

c. **12:40, 12:45, 12:50,** ______ ______ ______

d. 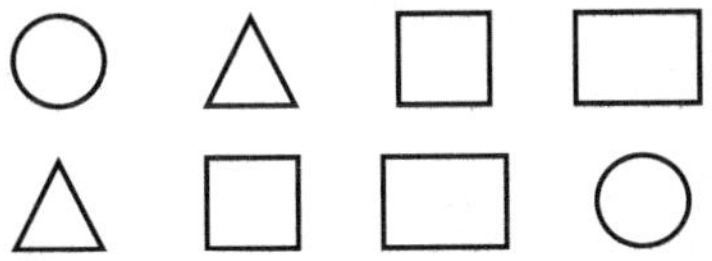 ______ ______ ______

2.

Materials: 16 counters (4 each of red, green, blue, and yellow)

Place counters on the grid, one to each square, so that no two identical colors are in the same row, column, or major diagonal.

Color the squares when you have completed the task.

MORE NUMBER PATTERN PRACTICE

Unit 2

Completing patterns
Shape patterns
Counting
Creating
Following rules

Objectives

- *generate and describe number patterns using a variety of strategies*
- *check whether answers to questions are correct and sensible*
- *appreciate the importance of visualization when solving problems*
- *appreciate that mathematics involves observing, generalizing, and representing patterns and relationships*
- *predict or approximate the value of a given term in a sequence*

Language

patterns, sequence, shapes, puzzles

Materials/Resources

colored pencils, blank paper

Contents of Student Pages

* *Materials needed for each reproducible student page*

Remember

Before starting ensure each student:

- ❑ *reads problem solving questions twice.*
- ❑ *develops the habit of checking.*
- ❑ *is encouraged to ask for help when needed.*

Additional Activities

- *Students use colored paper and fold it over 6 times. Then cut out shapes to make a pattern chain.*

  Examples: 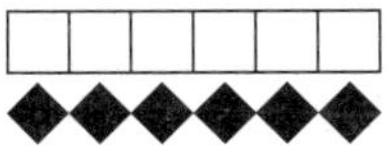

- *Students study the patterns that help with times tables.*
- *Students make large number lines and step on number patterns. This could be done with chalk on the pavement (with school's permission).*

Answers

Page 104 Completing Patterns

1. a. 48, 96, 192 d. 750, 760, 770
   b. 31, 38, 46 e. 55, 40, 25
   c. 90, 105, 120 f. 44, 55, 66
2. a. +12: 12 → 24 → 36 → 48 → 60 → 72
   b. +20: 6 → 26 → 46 → 66 → 86 → 106
   c. −9: 90 → 81 → 72 → 63 → 54 → 45
   d. +25: 10 → 35 → 60 → 85 → 110 → 135
   e. +125: 50 → 175 → 300 → 425 → 550 → 675
3. a.

●	16	12	21	7	35
x2	32	24	42	14	70
+15	31	27	36	22	50

b.

■	100	56	25	32	40
−12	88	44	13	20	28
+15	115	71	40	47	55

c.

★	14	28	33	16	18
−9	5	19	24	7	9
+22	36	50	55	38	40

d.

♥	42	16	22	17	8
+36	78	52	58	53	44
+13	55	29	35	30	21

4. a. square, square, triangle, square
   b. circle, triangle, circle, square
   c. diamond, vertical rectangle, square, circle
   d. heart, two horizontal rectangles, heart, heart
   e. star, star, star, circle
   f. checkmark, checkmark, cross, checkmark

Page 105 Shape Patterns

Teacher to check individual work.

Page 106 Counting Patterns

1. a. 10, 15, 35, 50 c. 8, 12, 44, 80
   b. 8, 12, 36, 48
2. a. 1st Row: 26, 30, 34
   2nd Row: 39, 41, 43
   3rd Row: 38, 44, 48
   4th Row: 47, 51, 55, 57
   Last Row: 50, 56, 58, 62
   b. 1st Row: 16, 128, 256
   2nd Row: 10, 40, 160, 320
   3rd Row: 36, 72, 288, 576
   4th Row: 17, 136, 272, 544, 1088
   Last Row: 12, 48, 192, 384, 768
3. Teacher to check individual work.

Page 107 Grid Patterns

Teacher to check individual work.

Page 108 Following Rules

1. a. 40, 84, 16, 68, 24, 52
   b. 26, 70, 2, 54, 10, 38
   c. 49, 115, 13, 91, 25, 67
   d. 10, 21, 4, 17, 6, 13
   e. 26 ,70, 2, 54, 10, 38
2. 68, 71, 97, 53, 40, 16, 37, 28, 45
3. 170, 161, 152, 145, 137, 127, 115, 144, 131
4. a. 19, 23, 27, 31 e. 29, 23, 17, 11
   b. 72, 63, 54, 45 f. 10, 14, 12, 16
   c. 35, 44, 53, 62 g. 23, 28, 27, 32
   d. 17, 22, 27, 32 h. 60, 65, 70, 75
5. Teacher to check individual work.

Page 109 Number Line Patterns

1. a. Add 2
   b. Add 1, Add 1, Add 2
   c. Take 4, Add 2
   d. Take 3, Take 2
2. a. 16, 22, 28
   b. 65, 55, 45
   c. 50, 500, 5,000
   d. 32, 16, 8
   e. 49, 65, 81
   f. 12, 48, 192
   g. 90, 70, 50
   h. 24, 12, 6
3. Teacher to check individual work.

Page 110 Assessment

1. a. 9, 13, 12, 16
   b. 80, 160, 320, 640
   c. 26, 34, 42, 50
   d. 81, 87, 93, 99
   e. 36, 30, 23, 15
   f. 22, 31, 32, 41
2. a. 36, 46, 56, 66, 76
   b. 83, 66, 49, 32, 15
   c. 4, 8, 16, 32, 64
   d. 32, 16, 8, 4, 2
3. Teacher to check individual work.
4. Teacher to check individual work.
5. a. 56, 36, 84, 32, 48, 68
   b. 6, 1, 13, 0, 4, 9
   c. 76, 46, 118, 40, 64, 94

Name **Date**

1. Complete these number patterns.

a. 6, 12, 24, ______ , ______ , ______
b. 16, 20, 25, ______ , ______ , ______
c. 45, 60, 75, ______ , ______ , ______
d. 720, 730, 740, ______ , ______ , ______
e. 100, 85, 70, ______ , ______ , ______
f. 11, 22, 33, ______ , ______ , ______

2. Use the rule in the square to fill in the missing numbers.

a. [+12] 12 → ___ → ___ → ___ → ___ → ___
b. [+20] 6 → ___ → ___ → ___ → ___ → ___
c. [–9] 90 → ___ → ___ → ___ → ___ → ___
d. [+25] 10 → ___ → ___ → ___ → ___ → ___
e. [+125] 50 → ___ → ___ → ___ → ___ → ___

3. Use the rules to complete the patterns.

a.

●	16	12	21	7	35
x2					
+15					

b.

■	100	56	25	32	40
–12					
+15					

c.

★	14	28	33	16	18
–9					
+22					

d.

♥	42	16	22	17	8
+36					
+13					

4. Complete each pattern of shapes.

a. △, □, □, △, ___, ___, ___, ___
b. ○, △, ○, □, ___, ___, ___, ___
c. ◇, ▯, □, ○, ___, ___, ___, ___
d. ♡, ♡, ⊟, ♡, ___, ___, ___, ___
e. ☆, ☆, ☆, ○, ___, ___, ___, ___
f. ✖, ✓, ✓, ✖, ___, ___, ___, ___

Name **Date**

1. Complete each of the shapes so that they match the first shape.

a.

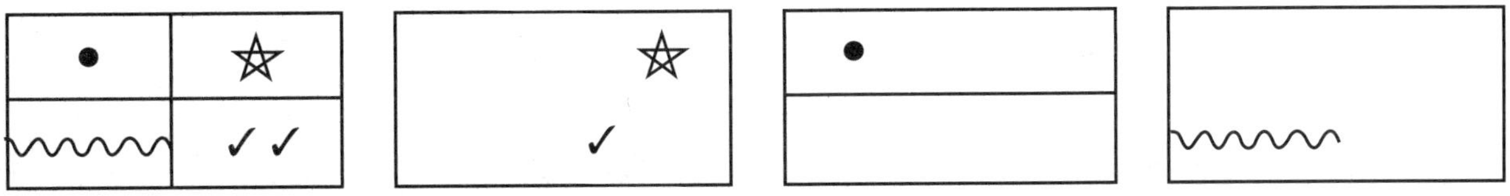

b.

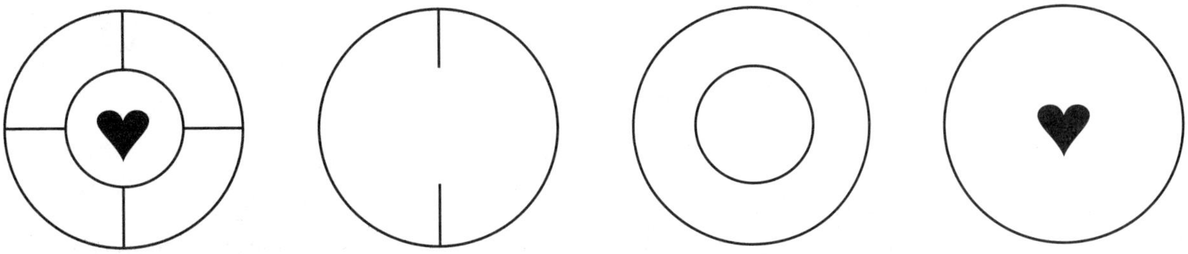

c.

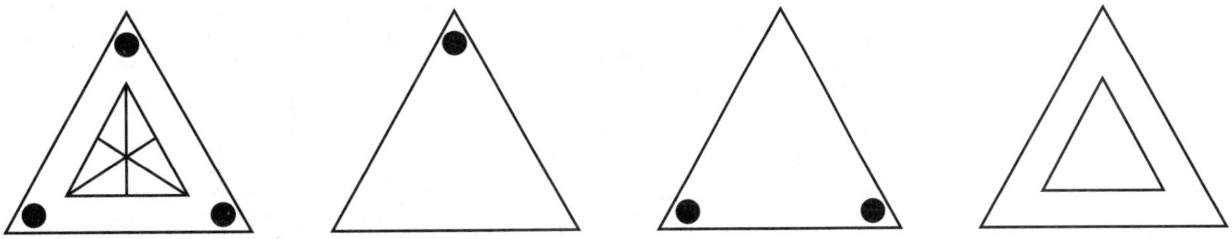

d.

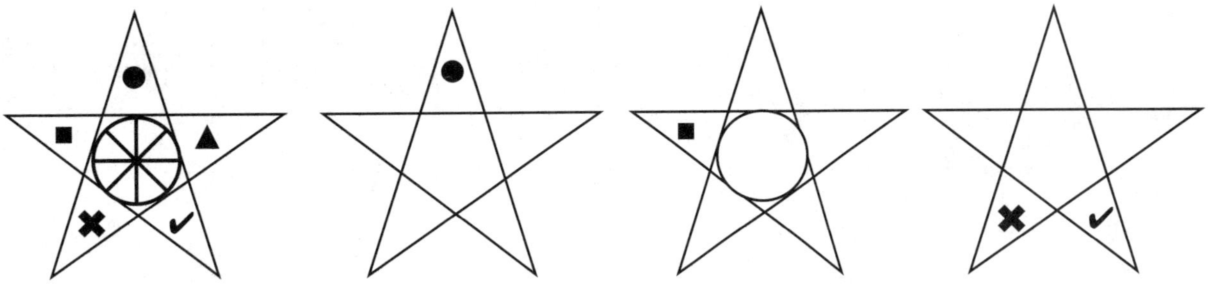

e.

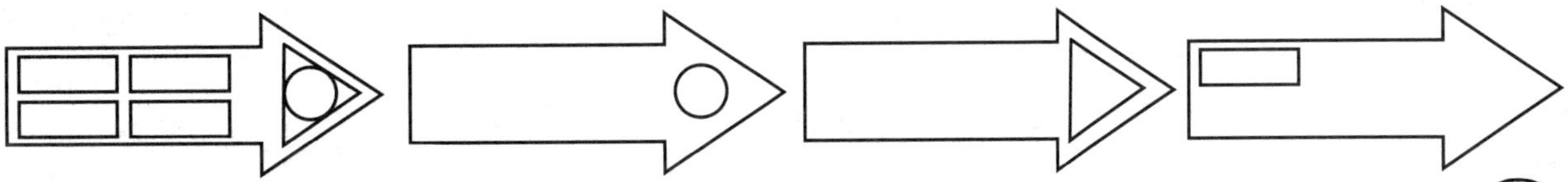

Name **Date**

1. Look at each pattern and complete the tables.

a. 

Number of stars	1	2	3	7	10
Number of triangles	5				

b. 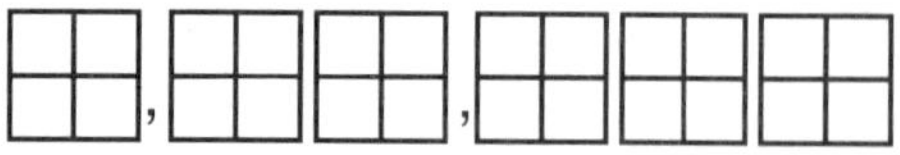

Number of large squares	1	2	3	9	12
Number of small squares	4				

c. 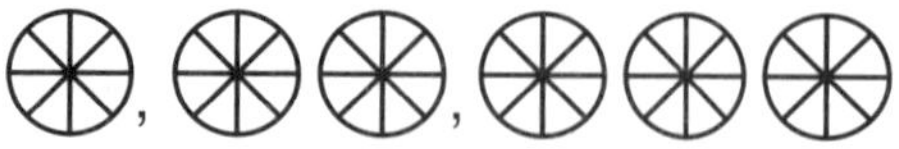

Number of circles	1	2	3	11	20
Number of lines	4				

2. Look for the pattern in the tables. Fill in the missing numbers.

a.

22	24		28		32	
33	35	37				45
36		40	42		46	
	49		53			59
	52	54			60	

b.

4	8		32	64		
5		20		80		
9	18			144		
	34	68				
	24		96			

3. Use colors to create a pattern in the boxes below.

Name **Date**

1. Some squares on the grid have been shaded. Finish shading the squares to continue the pattern.

a.

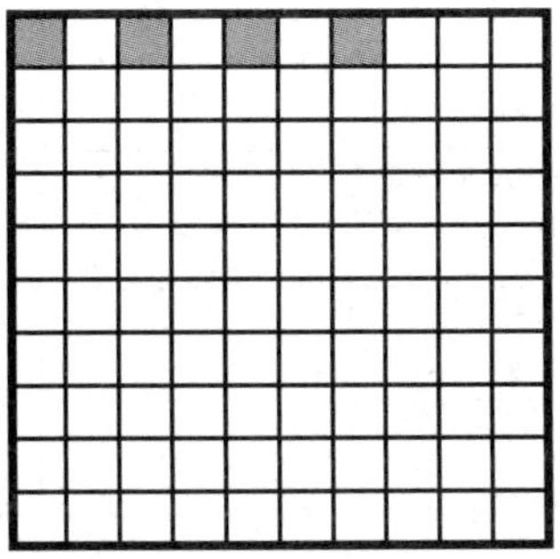

b.

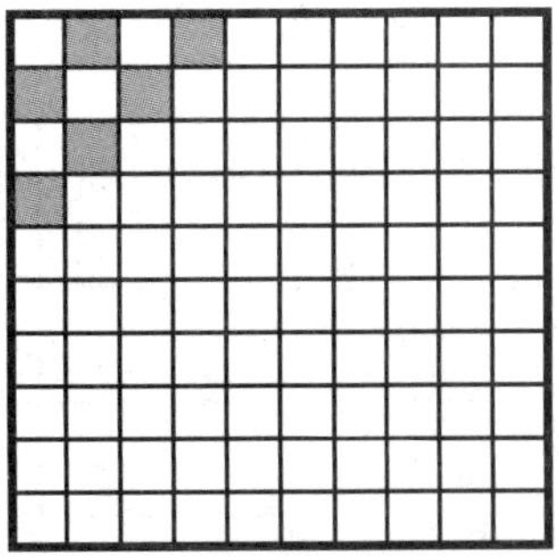

c.

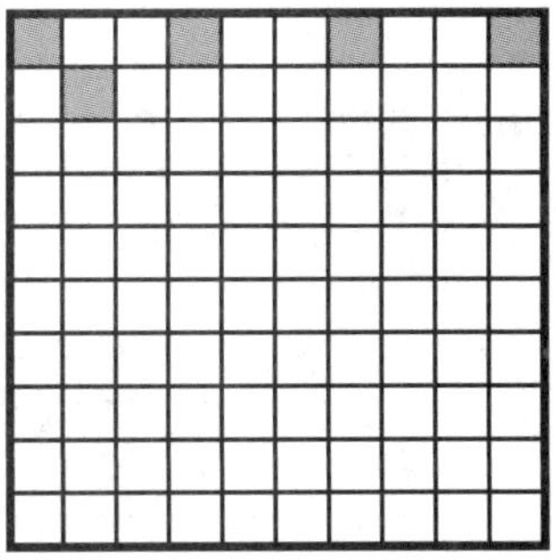

2. Read the rule below the square. Color the squares to make the pattern.

a.

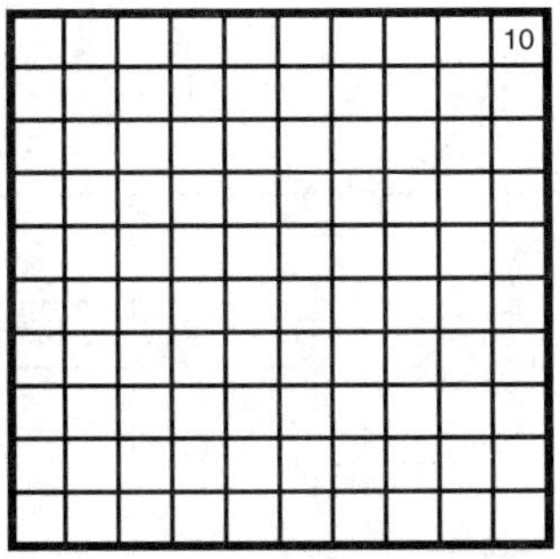

Start at 10 and add 10.

b.

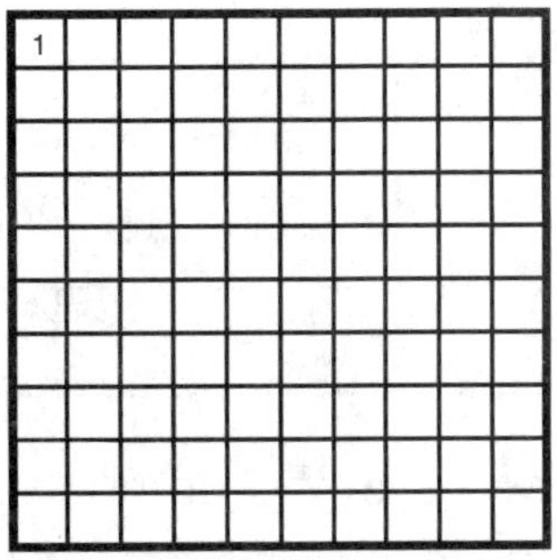

Start at 1 and add 5.

c.

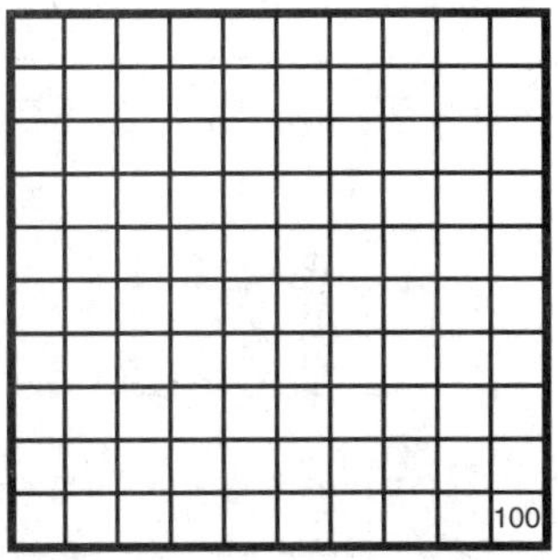

Start at 100 and add 5.

d.

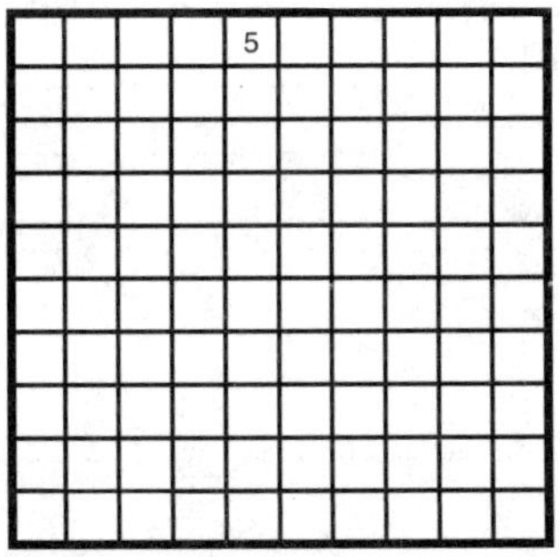

Start at 5 and add 15.

e.

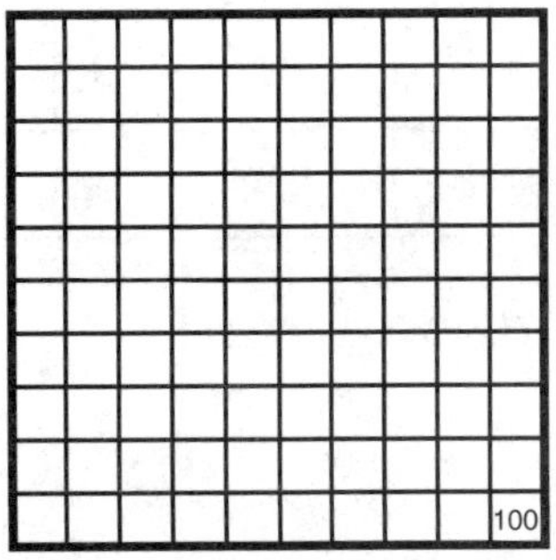

Start at 100 and subtract 1.

f.

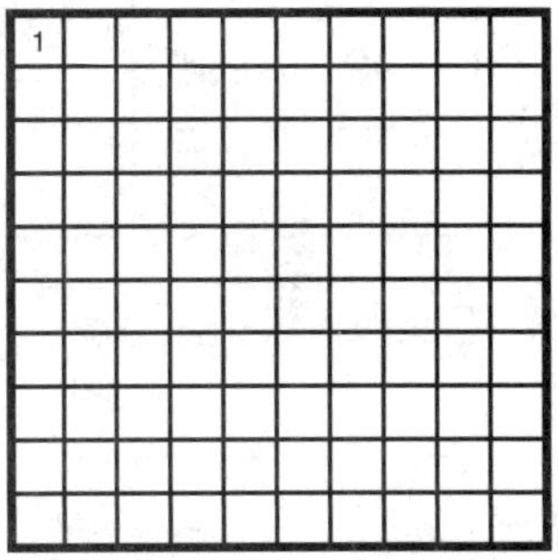

Start at 1 and add 25.

3. Make up your own rule and color the patterns.

a.

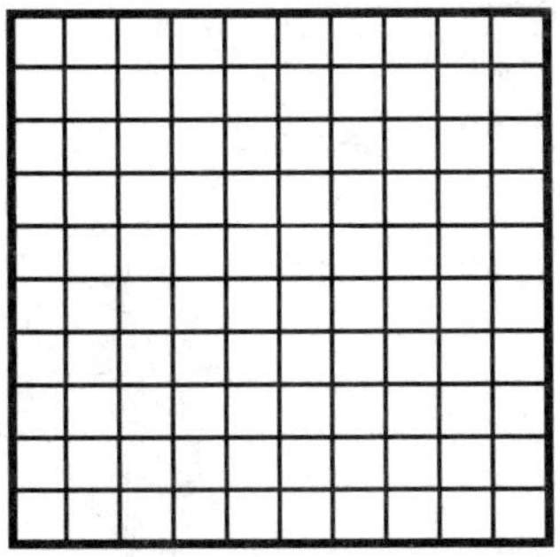

b.

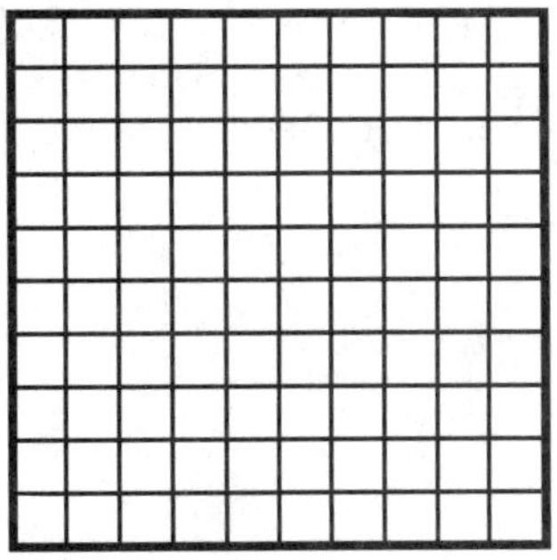

c.

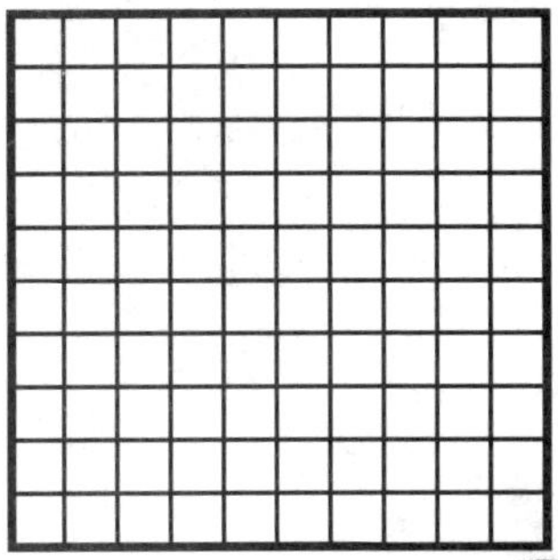

Name **Date**

1. Complete the table by following the rule.

	RULE	18	40	6	32	10	24
a.	Add 2; then double						
b.	Take 5 away; multiply by 2						
c.	Triple; take 5 away						
d.	Add 2; divide by 2						
e.	Double; take away 10						

2. Complete the boxes by adding 9 to the numbers.

59	62	88	44	31	7	28	19	36
+9	+9	+9	+9	+9	+9	+9	+9	+9

Did you find a quick way to add 9 to any number?
____
____
____

3. Complete the boxes by adding 99 to the numbers.

71	62	53	46	38	28	16	45	32
+99	+99	+99	+99	+99	+99	+99	+99	+99

Did you find a quick way to add 99 to any number?
____
____
____

4. Complete these number patterns.

a. 7, 11, 15, ___ , ___ , ___ , ___

b. 99, 90, 81, ___ , ___ , ___ , ___

c. 8, 17, 26, ___ , ___ , ___ , ___

d. 2, 7, 12, ___ , ___ , ___ , ___

e. 47, 41, 35, ___ , ___ , ___ , ___

f. 6, 10, 8, 12, ___ , ___ , ___ , ___

g. 15, 20, 19, 24, ___ , ___ , ___ , ___

h. 45, 50, 55, ___ , ___ , ___ , ___

5. Write some of your own number patterns. Ask a friend to solve them.

Name **Date**

1. Find the rule for each number line.

a.
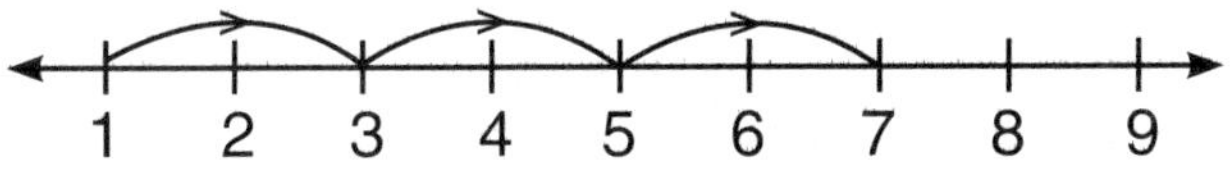

Rule: ____________________

b.
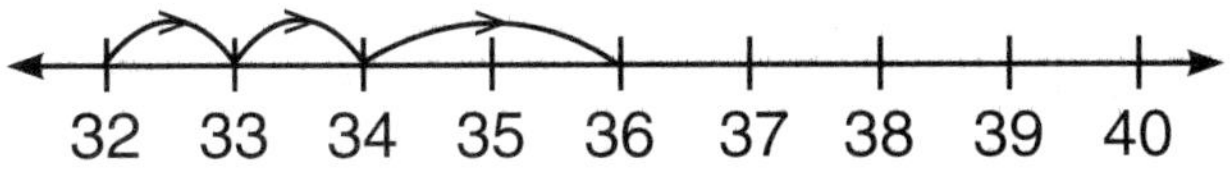

Rule: ____________________

c.
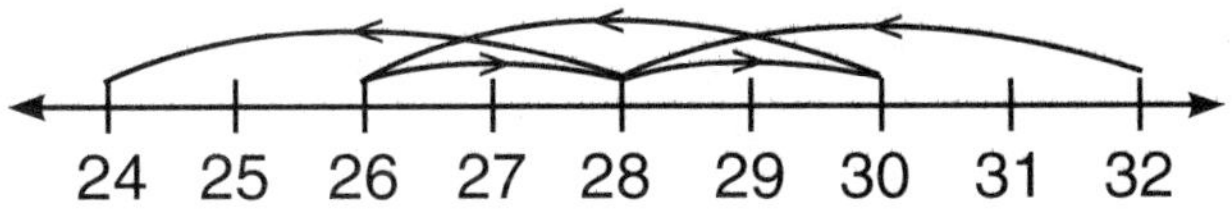

Rule: ____________________

d.
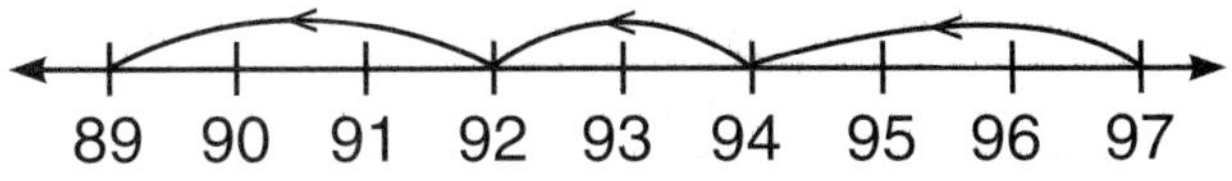

Rule: ____________________

2. Find the next three numbers.

a. 10; (add 6), ___ , ___ , ___

b. 75; (subtract 10), ___ , ___ , ___

c. 5; (multiply by 10), ___ , ___ , ___

d. 64; (divided by 2), ___ , ___ , ___

e. 33; (add 16), ___ , ___ , ___

f. 3; (multiply by 4), ___ , ___ , ___

g. 110; (take away 20), ___ , ___ , ___

h. 48; (divided by 2), ___ , ___ , ___

3. Complete these patterns.

a.
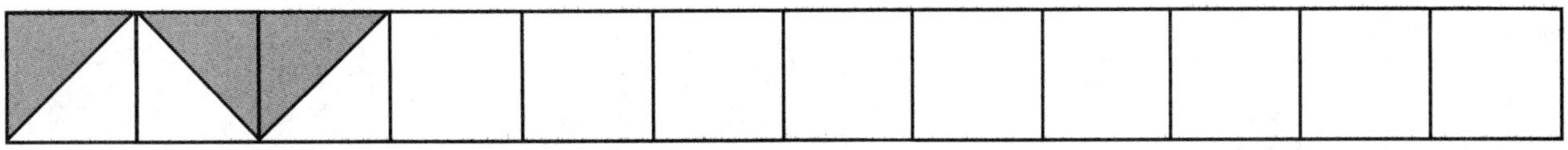

b.
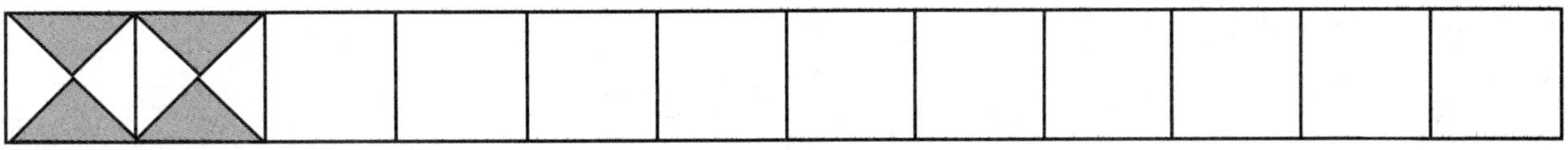

c.
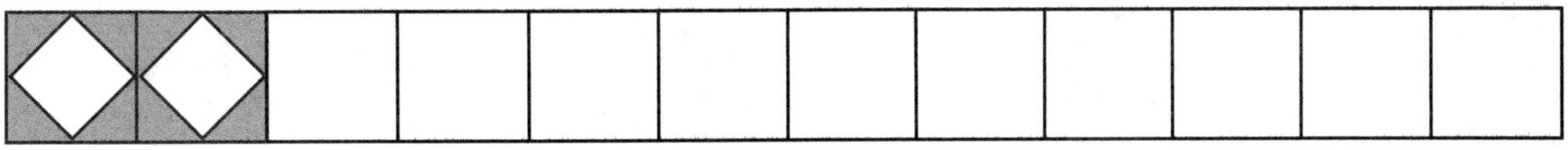

Name **Date**

1. Complete these number patterns.

a. 3, 7, 6, 10, ____ , ____ , ____ , ____

b. 10, 20, 40, ____ , ____ , ____ , ____

c. 2, 10, 18, ____ , ____ , ____ , ____

d. 63, 69, 75, ____ , ____ , ____ , ____

e. 50, 48, 45, 41, ____ , ____ , ____ , ____

f. 1, 2, 11, 12, 21, ____ , ____ , ____ , ____

2. Use the rule to complete the number pattern.

a. +10 | 26 → ( ) → ( ) → ( ) → ( ) → ( )

b. –17 | 100 → ( ) → ( ) → ( ) → ( ) → ( )

c. x2 | 2 → ( ) → ( ) → ( ) → ( ) → ( )

d. ÷2 | 64 → ( ) → ( ) → ( ) → ( ) → ( )

3. Complete the patterns.

a.

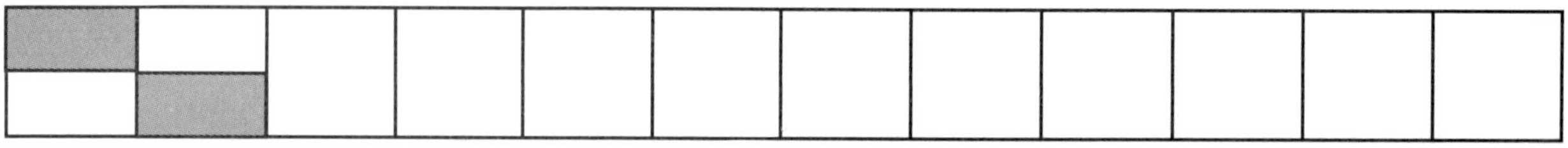

b.

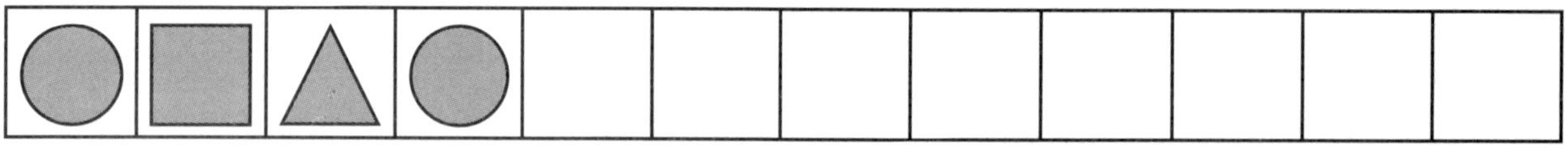

4. Read each rule and color the squares to make the pattern.

a.

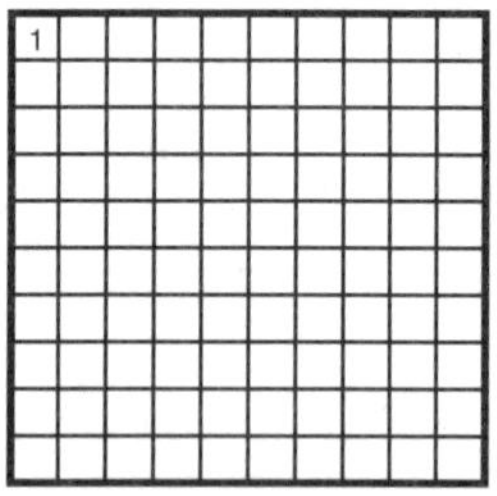

Start at 1 add 12.

b.

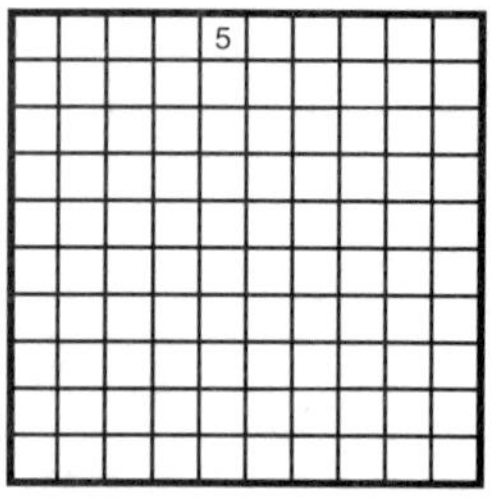

Start at 5 add 5.

c.

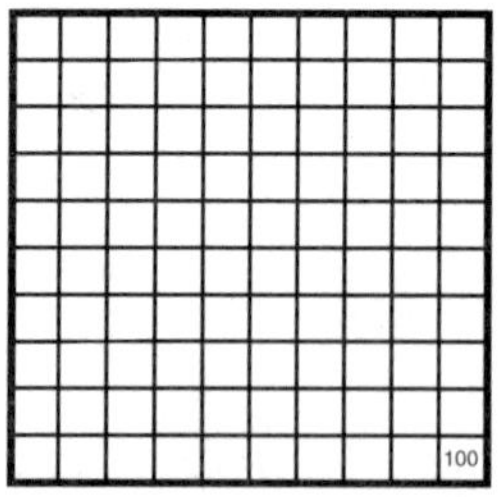

Start at 100 subtract 17.

5. Complete the table by following the rules.

	RULE	22	12	36	10	18	28
a.	add 6; double						
b.	divide by 2; take away 5						
c.	triple; add 10						

Skills Index

The following index lists specific objectives for the student pages of each unit in the book. The objectives are grouped according to the sections listed in the Table of Contents. Use the Skills Index as a resource for identifying the units and student pages you wish to use.

Division (cont.)

Number Facts

Number Patterns

